计算机课程设计与综合实践规划教材

编译原理实验教程

刘 刚 赵鹏翀 编著

清华大学出版社
北京

内容简介

本书结合对现代编译器设计理论的详细研究，精心设计了若干个实验，每个实验都使用C语言编写完成，并配有大量的练习，使读者可以将注意力集中到编程上，真正做到"以源代码为核心"。读者可以亲自动手完成这些实验，在实践的过程中循序渐进地学习编译原理的理论知识，进而加深对编译原理的理解，掌握理论知识在实际中的应用情况，从而将理论知识统一起来。本书还完整描述了一个可运行的小规模语言编译器（包括源代码）。

全书包含12个实验，是一本真正能够引导读者动手实践的书。本书可作为高等院校"编译原理"课程的实践教材，也可作为各类程序开发者、爱好者的阅读参考书。

本书封面贴有清华大学出版社防伪标签，无标签者不得销售。
版权所有，侵权必究。举报：010-62782989，beiqinquan@tup.tsinghua.edu.cn。

图书在版编目(CIP)数据

编译原理实验教程/刘刚，赵鹏翀编著. —北京：清华大学出版社，2017（2023.12重印）
（计算机课程设计与综合实践规划教材）
ISBN 978-7-302-45699-5

Ⅰ. ①编… Ⅱ. ①刘… ②赵… Ⅲ. ①编译程序－程序设计－高等学校－教材 Ⅳ. ①TP314

中国版本图书馆 CIP 数据核字（2016）第 288808 号

责任编辑：袁勤勇　王冰飞
封面设计：傅瑞学
责任校对：焦丽丽
责任印制：沈　露

出版发行：清华大学出版社
　　网　　址：https://www.tup.com.cn，https://www.wqxuetang.com
　　地　　址：北京清华大学学研大厦A座　　邮　编：100084
　　社 总 机：010-83470000　　　　　　　　邮　购：010-62786544
　　投稿与读者服务：010-62776969，c-service@tup.tsinghua.edu.cn
　　质 量 反 馈：010-62772015，zhiliang@tup.tsinghua.edu.cn
　　课 件 下 载：https://www.tup.com.cn，010-83470236
印 装 者：北京建宏印刷有限公司
经　　销：全国新华书店
开　　本：185mm×260mm　　印　张：14.75　　字　数：337千字
版　　次：2017年3月第1版　　　　　　　　印　次：2023年12月第6次印刷
定　　价：46.00元

产品编号：072744-02

FOREWORD

前言

> 纸上得来终觉浅,绝知此事要躬行。
>
> ——陆游

本书特点

众所周知,编译原理是计算机知识领域中一个核心的组成部分,也是高校计算机科学专业学生的重要基础课。同时,编译原理也是一门实践性很强的课程。本书通过引导读者动手进行相应的实验,进而达到使读者深刻理解编译原理的目的。

本书非常适合于编译原理的初学者使用,能够帮助初学者进行高质量的编译系统实验。本书精心设计了若干个实验题目,使读者可以逐步地接触到一个实际编译系统的源代码。本书还配置了一个集成度很高的实验环境——CP Lab,在这个集成实验环境中,读者可以非常轻松地编辑、编译和调试源代码,从而读者可以将有限的精力放在学习编译原理上,而不是如何构建实验环境,或者使用各种工具上。本书会一步一步地带动读者通过动手实践的方式来分析和编写可以实际工作的源代码,进而理解编译原理。

现代编译系统已经变得越来越复杂,虽然本书中的编译系统是专为教学而设计,相对于一些商用编译系统已经非常简化,但是相信本书的很多读者都是第一次接触到这样规模的源代码。本书在编写时充分考虑到了这个问题,并做了一些有益的尝试。为了方便读者理解,编译系统的源代码进行了很多简化,各个模块间的耦合性被设计得尽量小,这样读者在学习某个模块时就更容易集中精力。在本书的各个实验之间也会存在一些交叉的或者重复的内容,有时还会提示读者回到之前的实验学习相关的内容,这种"螺旋式"的学习方法可以帮助读者从整体上理解编译原理。

本书另外一个着重点就是要让读者真正动手实践。只有通过亲身实践学习到的知识才能够真正被掌握,而那些仅仅从书本上得到的知识更容易被忘记。本书为了让读者在动手实践的过程中达到"做中学"的目的,精心设计了 12 个配套实验,可以覆盖编译原理知识领域中所有重要的模块和知识点。本书配套的实验按照"由易到难,循序渐进"的原则进行设计。前面的若干个实验以"验证型"练习为主,后面的若干个实验会添加适当的"设计型"和"综合型"练习。在单个实验内容的安排上,一般会首先带领读者阅读并调试相关模块的源代码,并结合相应的编译原理进行分析。待读者对源代码和系统原理熟悉后,再安排读者对已有代码进行适当改写,或者编写新的代码。在每个实验的最后还会提供一些"思考与练习"的题目,感兴趣的读者可以完成这些题目,从而进一步提高动手实践能力和创新能力。

此外,考虑到在实际工作中,如果一位刚参加工作的程序员参与到一个项目的开发中,项目负责人一定会让他首先阅读项目已有的代码,并在已有代码的基础上进行一些小的修改,待他工作一段时间后,对项目有较深入的理解,才能在项目中添加一些复杂的、创新的功能。读者按照本书提供的实验进行实践的过程,与上述过程是完全一致的,这也是本书实验设计的目的之一。

研究表明,图示具有直观、简洁、易于说明事物的客观现状或事件的发展过程的特点。在对某一事物或事件进行描述时,图示往往比文字更容易被读者理解和接受。所以,本书不遗余力地使用各种图示或者表格,力求将枯燥、复杂的编译原理,以更直观的方式展现在读者的面前。而且,本书在适当的地方会从源代码中引用一些关键的代码片段,并结合编译原理对这些代码片段进行详细讲解,让读者有一种身临其境的真实感。

阅读源代码的建议

接下来为读者阅读源代码提供一些有益的建议,希望能够帮助读者更顺利地阅读源代码。

首先,读者应该明确阅读源代码的目的,或者说通过阅读源代码,读者能够学到哪些有用的知识,对读者参加实际工作会有哪些帮助。最重要的目的当然是理解编译系统的原理,源代码能够帮助读者将书本上枯燥的理论实例化。虽然读者亲自动手开发一个商业编译系统的可能性很小,但是编译系统所使用的许多思想在计算机科学的各个领域有广泛的适用性,学习系统的内部设计理念对于算法设计和实现、构建虚拟环境、网络管理等其他多个领域也非常有用。而且,源代码是精心编写的高质量源代码,无论是代码的组织结构还是代码的编写风格,都是按照商业级的规格来完成的,这些在读者的实际工作中都会有很大的借鉴意义。此外,本书由于篇幅的限制,不可能涉及系统的所有内容,幸好源代码就是最完全、最准确的文档,读者通过学习源代码,能够获得几倍于本书内容的知识。

其次,读者在开始阅读源代码之前,还应该完成一些准备工作。源代码主要使用 C 语言编写,定义较多的数据结构,并尽量使用常用的、简单的算法来操作这些数据结构,所以读者需要有比较扎实的 C 语言程序设计、数据结构和算法的相关知识。

在阅读源代码时应该使用一些正确的方法,从而达到事半功倍的效果。已经有专门的书籍详细介绍阅读源代码的方法,本书由于篇幅的限制,在这里只能为读者列举一些快速而有效的方法。

- 应该首先搞清楚源代码的组织方式,例如都包含哪些源代码文件,这些源代码文件是如何组织在不同的文件夹中的。这对于读者快速地了解结构有很大帮助。
- 重视数据结构。要搞清楚数据结构中各个域的意义,以及使用这些数据结构定义了哪些重要的变量(特别是全局变量)。系统的大部分函数都是在操作由这些数据结构所定义的变量,要搞清楚函数对这些变量进行的操作会产生怎样的结果。
- 分析函数的层次和调用关系。要特别注意哪些函数是全局函数,哪些函数是模块内部使用的函数。
- 本书对于特别简单或者特别复杂的函数会一语带过,读者也可以在搞清楚这些函数功能的基础上暂时跳过它们,从而将有限的时间和精力用于学习本书详细介绍的重要函数。
- 重视阅读源代码文件中的注释,必要的情况下可以根据自己的理解添加一些注释。
- 使用工具提高阅读源代码的效率。
- 每当阅读完一部分源代码后,应该认真思考一下,大胆地提出一些问题,例如"为什么要这样编写?可以不可以用别的方法来编写?"也可以试着向别人介绍自己正在阅读的源代码,或者将自己的心得发布到互联网上。以某种方式表达自己思想的过程,其实就是重新梳理自己知识的过程,这样能够让读者的知识更加系统化,并且有可能发现被忽略掉的细节。
- 读者除了可以完成本书配套的实验之外,还可以自己设计一些小实验,例如进行一些修改或者添加一些功能来验证读者的想法。

参与讨论

读者可以使用下面的链接登录本书配套的论坛。论坛中有和读者一样对编译原理感兴趣的网友,有本书的编者,还有 CP Lab 的开发者。读者在这里提出的问题可以获得及时准确的解答,提出的意见和建议也可以在本书的下一版中获得虚心的接纳。如果本书有勘误信息或者更新的章节内容,也会在第一时间发布到论坛上。可以说,有一个高效的团队在为本书的读者服务,读者在使用本书学习的过程中可以获得持续的支持。

http://www.engintime.com/forum/

本书由哈尔滨工程大学刘刚、北京英真时代科技有限公司赵鹏翀编著,参编人员包括北京英真时代科技有限公司刘建成等。全书由刘刚进行统稿。

本书在撰写过程中参阅了大量的文献及资料,特此对这些作者表示诚挚的敬意。

编译技术的发展日新月异,新的技术也不断出现。由于时间仓促及作者视野的限制,书中难免出现疏漏及不当之处,敬请广大读者批评指正。

编 者
2017 年 1 月

目 录

CP Lab 简介 ·· 1

实验 1　实验环境的使用 ··· 3

实验 2　NFA 到 DFA ·· 26

实验 3　使用 Lex 自动生成扫描程序 ··· 48

实验 4　消除左递归(无替换) ·· 62

实验 5　消除左递归(有替换) ·· 74

实验 6　提取左因子 ··· 89

实验 7　First 集合 ··· 104

实验 8　Follow 集合 ·· 117

实验 9　Yacc 分析程序生成器 ·· 132

实验 10　符号表的构建和使用 ··· 137

实验 11　三地址码转换为 P-代码 ··· 146

实验 12　GCC 编译器案例综合研究 ·· 156

附录 A　TINY 编译器和 TM 机 ·· 167

参考文献 ·· 225

CP Lab 简介

> 获得人生中的成功需要的专注与坚持不懈多过天才与机会。
> ——C. W. Wendte

一、概述

CP Lab 是一款专用于高等院校编译原理实验教学的集成环境,具有可视化程度高、操作简便、扩展性强等特点。CP Lab 主要由两部分组成:

- 一个功能强大的 IDE 环境。
- 一套精心设计的编译原理实验题目和配套源代码。

CP Lab 所提供的 IDE 环境可以在 Windows 操作系统上快速安装、卸载,其用户界面和操作方式与 Microsoft Visual Studio 完全类似,有经验的读者可以迅速掌握其基本用法。该 IDE 环境提供的强大功能可以帮助读者顺利地完成编译原理实验,主要功能包括对实验源代码的演示功能、编辑功能、编译功能、调试功能和验证功能。

CP Lab 提供了一套精心设计的编译原理实验题目和配套源代码。实验题目涵盖了词法分析、语法分析、语义分析、代码生成等所有重要的编译原理和算法。完成这些实验题目后,读者可以学习到通过手工编码是如何一步步实现一个编译器的,还可以使用 Lex、Yacc 工具完成简单的词法分析程序和语法分析程序,甚至还可以深入分析一个小型的开源编译器的实现方法。所有的源代码文件都使用 C 语言编写,可以与主流编译原理教材配套使用。这些源代码以模块化的方式进行组织,并配有完善的中文注释,可读性好,完全符合商业级的编码规范。

使用 CP Lab 进行编译原理实验的过程如图 0-1 所示。

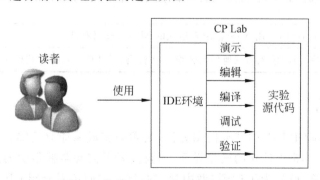

图 0-1　使用 CP Lab 进行编译原理实验

二、实验题目清单

下面的实验题目适合在学习编译原理课程的过程中逐个完成。

序号	概念	实验名称	难度	学时
1	词法分析	实验环境的使用（正则表达式到 NFA）	★★☆☆	2
2		NFA 到 DFA	★★★★	4
3		使用 Lex 自动生成扫描程序	★★☆☆	2
4	自顶向下的语法分析	消除左递归（无替换）	★★★☆	2
5		消除左递归（有替换）	★★★★	4
6		提取左因子	★★★☆	2
7		First 集合	★★☆☆	2
8		Follow 集合	★★☆☆	2
9	自底向上的语法分析	Yacc 分析程序生成器	★★☆☆	2
10	语义分析	符号表的构建与使用	★★☆☆	2
11	代码生成	三地址码转换为 P-代码	★★☆☆	2
12		GCC 编译器案例综合研究	★★☆☆	2

下面的实验题目适合在学完编译原理课程后，选择一个作为课程设计题目来完成。

序号	课程设计题目要求
1	编写一个程序，可以根据输入的正则表达式生成 NFA，然后将 NFA 转换为最小化的 DFA，最后使用得到的 DFA 完成字符串匹配。
2	参考 Lex 生成 TINY 语言扫描程序的过程，使用 Lex 为 C-Minus 语言生成一个扫描程序。
3	编写一个程序，可以为输入的 BNF 消除左递归或提取左因子，然后根据 BNF 计算出 First 集合和 Follow 集合，从而构造一个 LL(1) 分析表，最终实现一个表驱动的 LL(1) 分析算法。
4	参考 Yacc 生成 TINY 语言语法分析程序的过程，使用 Yacc 为 C-Minus 语言生成一个语法分析程序。
5	根据表达式的 BNF，使用 Yacc 输出表达式的 Modula-2 转换式。
6	编写一个程序，可将输入的三地址码转换为 P-代码，还可将输入的 P-代码转换为三地址码。

三、开始使用

读者迅速掌握 CP Lab 的基本使用方法，是顺利完成编译原理实验的重要前提。虽然 CP Lab 提供了非常人性化的操作界面，而且读者只需要掌握很少的几个核心功能就可以顺利完成实验，但是为了达到"做中学"的目的，在向读者介绍 CP Lab 的核心功能时，不会使用堆砌大量枯燥文字的方法，而是在后面的"实验 1"中，引导读者在使用 CP Lab 的过程中逐步掌握其基本使用方法，这样达到的效果最好。

EXPERIMENT 1

实验 1　实验环境的使用

实验难度：★★☆☆☆
建议学时：2 学时

一、实验目的

- 熟悉编译原理集成实验环境 CP Lab 的基本使用方法。
- 掌握正则表达式和 NFA 的含义。
- 实现正则表达式到 NFA 的转换。

二、预备知识

- 在这个实验中 NFA 状态结构体使用了类似于二叉树的数据结构，还包括了单链表插入操作以及栈的一些基本操作。如果读者对这一部分知识有遗忘，可以复习一下数据结构中的相关内容。
- 实验中需要把正则表达式转换为 NFA，所以要对正则表达式和 NFA（非确定有穷自动机）有初步的理解。读者可以参考配套的《编译原理》教材，预习这一部分内容。

三、实验内容

请读者按照下面的步骤完成实验内容，同时，仔细体会 CP Lab 的基本使用方法。在本实验题目中，操作步骤会编写得尽量详细，并会对 CP Lab 的核心功能进行具体说明。但是，在后面的实验题目中会尽量省略这些内容，而将重点放在实验相关的源代码上。如有必要，读者可以回到本实验题目中，参考 CP Lab 的基本使用方法。

3.1　启动 CP Lab

在安装有 CP Lab 的计算机上，可以使用两种不同的方法来启动 CP Lab：

- 在桌面上双击 Engintime CP Lab 图标。
- 单击"开始"菜单，在"程序"中的 Engintime CP Lab 中选择 Engintime CP Lab。

3.2　注册用户并登录

CP Lab 每次启动后都会弹出一个"登录"对话框，可以进行以下操作：

- 使用已有用户进行登录。

读者可以在"登录"对话框中填写已有的学号、姓名、密码完成登录。登录成功后，CP Lab 的标题栏会显示出读者用来登录的学号和姓名。

- 注册新用户。

读者可以单击"注册"按钮,在弹出的"注册"窗口中填写基本信息、所属机构、密码、密保问题完成注册,并自动登录。

3.3 主窗口布局

CP Lab 的主窗口布局由下面的若干元素组成:
- 顶部的菜单栏、工具栏。
- 停靠在左侧和底部的各种工具窗口。
- 余下的区域用来放置"起始页"和"源代码编辑器"窗口。

提示:菜单栏、工具栏和各种工具窗口的位置可以随意拖动。如果想恢复窗口的默认布局,选择"窗口"菜单中的"重置窗口布局"即可。

3.4 新建实验项目

新建一个实验项目的步骤如下:

(1) 在"文件"菜单中选择"新建"|"项目",打开"新建项目"对话框。

(2) 在"新建项目"对话框中选择项目模板"001 正则表达式到 NFA"。注意,其他模板会在后面的实验题目中使用。

(3) 在"名称"中输入新项目使用的文件夹名称"lab1"。

(4) 在"位置"中输入新项目保存在磁盘上的位置"C:\cplab"。

(5) 单击"确定"按钮。

新建完毕后,CP Lab 会自动打开这个新建的项目。在"项目管理器"窗口中(如图 1-1 所示),根结点是项目结点,各个子结点是项目包含的文件夹或者文件。读者也可以使用"Windows 资源管理器"打开磁盘上的"C:\cplab\lab1"文件夹,查看项目中包含的源代码文件。

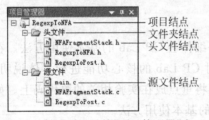

图 1-1 打开项目后的"项目管理器"窗口

提示:右击"项目管理器"窗口中的项目结点,选择快捷菜单中的"打开所在的文件夹",即可使用"Windows 资源管理器"打开项目所在的文件夹。

3.5 绑定课时

绑定课时功能可以帮助读者将新建的项目与一个课时完成绑定,操作步骤如下:

(1) 新建项目后,会自动弹出"绑定课时"对话框。

(2) 在"绑定课时"对话框中,读者可以根据课时信息选择一个合适的课时完成绑定。

(3) 绑定课时后,CP Lab 的标题栏会显示当前项目所绑定的课时信息。

注意:如果不为项目绑定课时,将会影响到实验课考评成绩,并且不能提交作业。在未登录的情况下,不会弹出"绑定课时"对话框。

3.6 阅读实验源代码

该实验包含了 3 个头文件 RegexpToNFA.h、RegexpToPost.h、NFAFragmentStack.h 和 3 个 C 源文件 main.c、RegexpToPost.c、NFAFragmentStack.c。下面对这些文件的主要内容、结构和作用进行说明。

1) main.c 文件(参见源代码清单 1-1)

在"项目管理器"窗口中双击 main.c 打开此文件。此文件主要包含以下内容：

（1）在文件的开始位置，使用预处理命令包含了 RegexpToNFA.h、RegexpToPost.h 和 NFAFragmentStack.h 文件。

（2）定义了 main 函数，在其中实现了栈的初始化；然后，调用了 re2post 函数，将正则表达式转换到解析树的后序序列；最后，调用了 post2nfa 函数，将解析树的后序序列转换到 NFA。

（3）在 main 函数的后面定义了函数 CreateNFAState 和 MakeNFAFragment，这两个函数分别用来创建一个新的 NFA 状态和构造一个新的 Fragment。接着定义了函数 post2nfa，关于此函数的功能、参数和返回值，可以参见其注释。在这个函数中用 $ 表示空转换，此函数的函数体还不完整，留给读者完成。

2) RegexpToPost.c 文件(参见源代码清单 1-2)

在"项目管理器"窗口中双击 RegexpToPost.c 打开此文件。此文件主要包含以下内容：

（1）在文件的开始位置，使用预处理命令包含了 RegexpToPost.h 文件。

（2）定义了 re2post 函数，此函数的主要功能是将正则表达式转换为解析树的后序序列形式。

3) NFAFragmentStack.c 文件(参见源代码清单 1-3)

在"项目管理器"窗口中双击 NFAFragmentStack.c 打开此文件。此文件主要包含以下内容：

（1）在文件的开始位置，使用预处理命令包含了 NFAFragmentStack.h 文件。

（2）定义了与栈相关的操作函数。在构造 NFA 的过程中，这个栈主要用来放置 NFA 片段。

4) RegexpToNFA.h 文件(参见源代码清单 1-4)

在"项目管理器"窗口中双击 RegexpToNFA.h 打开此文件。此文件主要包含以下内容：

（1）用到的 C 标准库头文件。目前只包含了标准输入输出头文件 stdio.h。

（2）其他模块的头文件。目前没有其他模块的头文件需要被包含。

（3）定义了与 NFA 相关的数据结构，包括 NFA 状态 NFAState 和 NFA 片段 NFAFragment。具体内容参见表 1-1 和表 1-2。

表 1-1

NFAState 的域	说　　明
Transform	状态间转换的标识。用 $ 表示 ε-转换
Next 1 和 Next 2	用于指向下一个状态。由于一个 NFA 状态可以存在多个转换,而在本程序中使用的是类似于二叉树的存储结构,每一个状态最多只有两个转换,所以,这里定义两个指针就足够了。当 NFA 只有一个转换时,优先使用 Next1,Next2 赋值为 NULL
Name	状态名称。使用整数表示(从 1 开始),根据调用 CreateNFAState 函数的顺序依次增加
AcceptFlag	是否为接受状态的标志。1 表示是接受状态 0 表示非接受状态

表 1-2

NFAFragment 的域	说　　明
StartState	NFAFragment 的开始状态
AcceptState	NFAFragment 的接受状态。在构造 NFA 的过程中总是在 NFA 的开始状态和接受状态上进行操作,所以用开始状态和接受状态表示一个 NFA 片段就足够了

(4) 声明函数和全局变量。

5) RegexpToPost.h 文件(参见源代码清单 1-5)

在"项目管理器"窗口中双击 RegexpToPost.h 打开此文件。此文件主要包含以下内容:

(1) 其他模块的头文件。目前只包含了头文件 RegexpToNFA.h。

(2) 声明函数。为了使程序模块化,将 re2post 函数声明包含在一个头文件中,再将此头文件包含在 main.c 中。

6) NFAFragmentStack.h 文件(参见源代码清单 1-6)

在"项目管理器"窗口中双击 NFAFragmentStack.h 打开此文件。此文件主要包含以下内容:

(1) 其他模块的头文件。目前只包含了头文件 RegexpToNFA.h。

(2) 定义重要的数据结构。定义了与栈相关的数据结构。

(3) 声明函数。声明了与栈相关的操作函数。

提示:请读者认真理解这部分内容,其他实验题目中的源代码文件也严格遵守这些约定,如无特殊情况将不再进行如此详细的说明。

源代码清单 1-1: main.c 文件

```
#include "RegexpToNFA.h"
#include "RegexpToPost.h"
#include "NFAFragmentStack.h"
```

```c
NFAFragmentStack FragmentStack;              //栈,用于存储 NFA 片段

char * regexp="ab";                          //例 1
//char * regexp="a|b";                       //例 2
//char * regexp="a*";                        //例 3
//char * regexp="a?";                        //例 4
//char * regexp="a+";                        //例 5
//char * regexp="a(a|1)*";                   //例 6
//char * regexp="(aa|b)*a(a|bb)*";           //例 7
//char * regexp="(a|b)*a(a|b)?";             //例 8

int main(int argc, char * * argv)
{
    char * post;
    NFAState * start;

    //
    //初始化栈
    //
    InitNFAFragmentStack(&FragmentStack);

    //
    //调用 re2post 函数将正则表达式字符串转换成解析树的后续遍历序列
    //
    post=re2post(regexp);

    //
    //调用 post2nfa 函数将解析树的后续遍历序列转换为 NFA,并返回开始状态
    //
    start=post2nfa(post);

    return 0;
}

/*
功能:
    初始化一个状态。

返回值:
    状态指针。
*/
int nstate=1;                                //状态名计数器
NFAState * CreateNFAState()
```

```
{
    NFAState * s=(NFAState * )malloc(sizeof(NFAState));

    s->Name=nstate++;
    s->Transform='\0';
    s->Next1=NULL;
    s->Next2=NULL;
    s->AcceptFlag=0;

    return s;
}
```

/*
功能：
    将开始状态和接受状态组成一个 Fragment。

参数：
    StartState--开始状态。
    AcceptState--接受状态。

返回值：
    Fragment 结构体指针。
*/

```
NFAFragment MakeNFAFragment(NFAState * StartState, NFAState * AcceptState)
{
    NFAFragment n={StartState, AcceptState};
    return n;
}
```

/*
功能：
    将解析树的后序序列转换为 NFA。

参数：
    postfix--解析树的后序序列指针。

返回值：
    NFA 的开始状态指针。
*/
```
const char VoidTrans='$ '; //表示空转换
NFAState * post2nfa(char * postfix)
{
    char * p;                                    //游标
```

```c
    NFAFragment fragment1, fragment2, fm;          //NFA 片段结构体变量
    NFAFragment fragment={0, 0};                   //初始化用于返回的 NFA 片段
    NFAState * NewStartState, * NewAcceptState;    //开始状态和接受状态指针

    //
                                                   //TODO: 在此添加代码
    //

    return fragment.StartState;
}
```

源代码清单 1-2：RegexpToPost.c 文件

```c
#include "RegexpToPost.h"

/*
功能：
    将输入的正则表达式字符串转换为解析树的后续遍历序列。

参数：
    re--正则表达式指针。

返回值：
    解析树的后续遍历序列数组指针。
*/
char * re2post(char * re)
{
    int nalt;              //表示解析到这个字符为止,已经有多少个分支结构
    int natom;             //表示解析到这个字符为止,已经有多少个原子结构
    static char buf[8000];
    char * dst;

    struct {
        int nalt;
        int natom;
    } paren[100], * p;

    p=paren;
    dst=buf;
    nalt=0;
    natom=0;
    if(strlen(re) >=sizeof buf/2)
        return NULL;
```

```
            for(; * re; re++){
                switch(* re){
                    case '(':
                        if(natom >1)
                        {
                            --natom;
                            * dst++='.';
                        }
                        if(p >=paren+100)
                            return NULL;
                        p->nalt=nalt;
                        p->natom=natom;
                        p++;
                        nalt=0;
                        natom=0;
                        break;
                    case '|':
                        if(natom==0)
                            return NULL;
                        while(--natom >0)
                            * dst++='.';
                        nalt++;
                        break;
                    case ')':
                        if(p==paren)
                            return NULL;
                        if(natom==0)
                            return NULL;
                        while(--natom >0)
                            * dst++='.';
                        for(; nalt >0; nalt--)
                            * dst++='|';
                        --p;
                        nalt=p->nalt;
                        natom=p->natom;
                        natom++;
                        break;
                    case '*':
                    case '+':
                    case '?':
                        if(natom==0)
                            return NULL;
                        * dst++= * re;
```

```
                break;
            default:
                if(natom >1)
                {
                    --natom;
                    *dst++='.';
                }
                *dst++=*re;
                natom++;
                break;
        }
    }
    if(p!=paren)
        return NULL;
    while(--natom >0)
        *dst++='.';
    for(; nalt >0; nalt--)
        *dst++='|';
    *dst=0;

    return buf;
}
```

源代码清单 1-3: **NFAFragmentStack.c** 文件

```
#include "NFAFragmentStack.h"

/*
功能：
    初始化栈。

参数：
    pS--栈的指针。
*/
void InitNFAFragmentStack(NFAFragmentStack *pS)
{
    pS->top=-1;
}

/*
功能：
    将元素入栈。

参数：
    pS--栈的指针。
```

        Elem--入栈的元素。

返回值:
    空。
*/
void PushNFAFragment(NFAFragmentStack * pS, NFAFragment Elem)
{
    //
    //栈满,入栈失败
    //
    if(MAX_STACK_LENGTH-1 <=pS->top)
        return;

    pS->top++;
    pS->buffer[pS->top]=Elem; //将元素插入栈顶

    return;
}

/*
功能:
    将栈顶元素出栈。

参数:
    pS--栈的指针。

返回值:
    如果出栈成功返回出栈元素的值。
    如果出栈失败返回-1。
*/
NFAFragment PopNFAFragment(NFAFragmentStack * pS)
{
    int pos;
    NFAFragment fragment={0, 0};

    //
    //栈为空,出栈失败
    //
    if(NFAFragmentStackEmpty(pS))
        return fragment;

    pos=pS->top;
    pS->top--;

```
        return pS->buffer[pos];
}


/*
功能：
    判断栈是否为空。

参数：
    pQ--栈的指针。

返回值：
    如果栈空返回 1(真)
    如果栈非空返回 0(假)
*/
int NFAFragmentStackEmpty(NFAFragmentStack * pS)
{
    return-1==pS->top ? 1 : 0;
}
```

源代码清单 1-4：RegexpToNFA.h 文件

```
#ifndef _REGEXPTONFA_H_
#define _REGEXPTONFA_H_

//
//在此处包含 C 标准库头文件
//

#include <stdio.h>

//
//在此处包含其他头文件
//


//
//在此处定义数据结构
//

typedef struct _NFAState{
    char Transform;              //状态间转换的标识,用'$ ' 表示'ε-转换'
    struct _NFAState * Next1;    //指向下一个状态
    struct _NFAState * Next2;    //指向下一个状态
```

```c
    int Name;                        //状态名称
    int AcceptFlag;     //是否为接受状态的标志,1表示是接受状态,0表示不是接受状态
}NFAState;

//Fragment 结构是一个 NFA 的片段
typedef struct _NFAFragment{
    NFAState * StartState;           //开始状态
    NFAState * AcceptState;          //接受状态
}NFAFragment;

//
//在此处声明函数
//

NFAState * CreateNFAState();
NFAState * post2nfa(char * postfix);
NFAFragment MakeNFAFragment(NFAState * StartState, NFAState * AcceptState);

//
//在此处声明全局变量
//

extern int nstate;
extern const char VoidTrans;
extern char * regexp;

#endif /* _REGEXPTONFA_H_ */
```

源代码清单 1-5：RegexpToPost.h 文件

```c
#ifndef _REGEXPTOPOST_H_
#define _REGEXPTOPOST_H_

//
//在此处包含 C 标准库头文件
//



//
//在此处包含其他头文件
//

#include "RegexpToNFA.h"
```

```
//
//在此处定义数据结构
//


//
//在此处声明函数
//

char * re2post(char * re);

#endif /* _REGEXPTOPOST_H_ */
```

源代码清单 1-6：NFAFragmentStack.h 文件

```
#ifndef _NFAFRAGMENTSTACK_H_
#define _NFAFRAGMENTSTACK_H_

//
//在此处包含C标准库头文件
//



//
//在此处包含其他头文件
//

#include "RegexpToNFA.h"

//
//在此处定义数据结构
//

#define MAX_STACK_LENGTH 1024          //栈的最大长度

//栈
typedef struct _NFAFragmentStack{
    NFAFragment buffer[MAX_STACK_LENGTH];    //栈的缓冲区
    int top;                                 //指示栈顶的位置,而不是栈中元素的个数
}NFAFragmentStack;

//
//在此处声明函数
//
```

```
void InitNFAFragmentStack(NFAFragmentStack * pS);
void PushNFAFragment(NFAFragmentStack * pS, NFAFragment Elem);
NFAFragment PopNFAFragment(NFAFragmentStack * pS);
int NFAFragmentStackEmpty(NFAFragmentStack * pS);

//
//在此处声明全局变量
//


#endif /* _NFAFRAGMENTSTACK_H_ */
```

3.7 生成项目

使用"生成项目"功能可以将程序的源代码文件编译为可执行的二进制文件,步骤如下:

(1) 在"生成"菜单中选择"生成项目"(快捷键是 F7)。

(2) 在项目生成过程中,"输出"窗口会实时显示生成的进度和结果。如果源代码中不包含语法错误,会在生成的最后阶段提示生成成功,如图 1-2 所示。

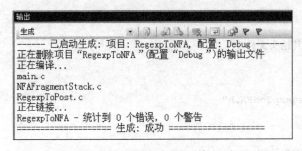

图 1-2　生成项目成功后的"输出"窗口

(3) 生成项目的过程,就是将项目所包含的每个 C 源代码文件(.c 文件)编译为一个对象文件(.o 文件),然后再将多个对象文件链接为一个目标文件(.exe 文件)的过程。以本实验为例,成功生成项目后,默认会在 C:\cplab\lab1\Debug 目录下生成 main.o 文件、RegexpToPost.o 文件、NFAFragmentStack.o 文件和 RegexpToNFA.exe 文件。

提示:读者可以通过修改项目名称的方法来修改生成的 .exe 文件的名称。方法是:在"项目管理器"窗口中右击项目结点,选择快捷菜单中的"重命名"命令。待项目名称修改后,需要再次生成项目才能得到新的 .exe 文件。

3.8 解决语法错误

如果在源代码中存在语法错误,在生成项目的过程中,"输出"窗口会显示相应的错误信息(包括错误所在文件的路径、错误在文件中的位置以及错误原因),并在生成的最后阶段提示生成失败。此时,在"输出"窗口中双击错误信息所在的行,CP Lab 会使用源代码编辑器自动打开错误所在的文件,并定位到错误所在的代码行。

可以按照下面的步骤进行练习:
(1) 在源代码文件中故意输入一些错误的代码(例如删除一个代码行结尾的分号)。
(2) 生成项目。
(3) 在"输出"窗口中双击错误信息来定位存在错误的代码行,并将代码修改正确。
(4) 重复步骤(2)、(3),直到项目生成成功。

3.9 观察点和演示模式

这里介绍 CP Lab 提供的两个重要功能:**观察点**和**演示模式**。

1. 观察点

一个观察点对应一个函数的起始位置和结束位置(称这个函数为观察点函数)。在调试过程中,当程序执行到观察点函数的起始位置和结束位置时就会发生中断,就好像在这两个位置上添加了断点一样。并且,只要在观察点函数内部发生中断(包括命中断点、单步调试等),就会在"转储信息"窗口中显示观察点函数正在操作的数据信息,如果在"演示模式"下,还会在"演示流程"窗口中显示观察点函数的流程信息。

以本实验为例,"观察点"窗口如图 1-3 所示(在"调试"菜单的"窗口"中选择"观察点",可以打开"观察点"窗口),说明 post2nfa 函数是一个观察点函数。启动调试后,在 main.c 文件 post2nfa 函数的开始位置和结束位置的左侧空白处会显示观察点图标(与"观察点"窗口中左侧的图标一致),当程序执行到 post2nfa 函数的开始位置和结束位置时会发生中断。启动调试后,观察点窗口如图 1-4 所示,可以显示出观察点所在的"文件"和"地址"。

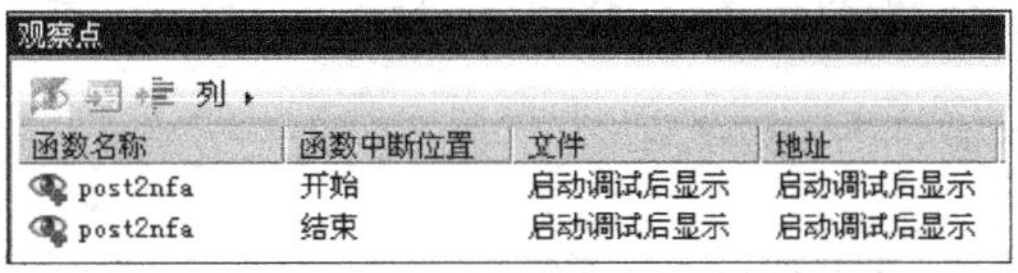

图 1-3 观察点窗口(未启动调试)

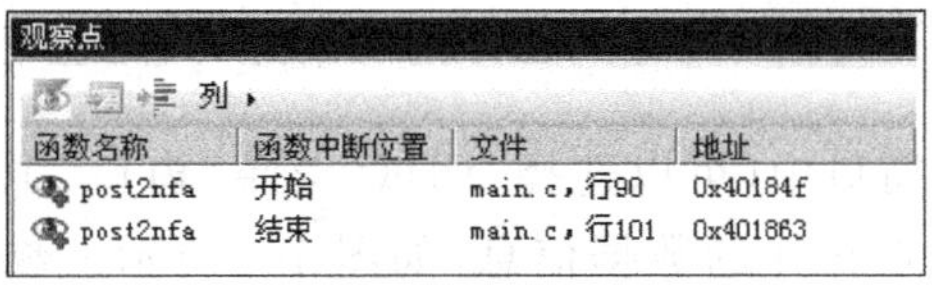

图 1-4 观察点窗口(启动调试)

2. 演示模式

当 CP Lab 工具栏上的"演示模式"按钮高亮显示时(如图 1-5 所示),CP Lab 处于演示模式。当在演示模式下调试观察点函数时,会忽略掉其函数体中的所有代码和断点,取而代之的是使用 CP Lab 提供的演示功能对观察点函数的执行过程和返回值进行演示。此特性可使观察点函数在还未完整实现的情况下,让读者了解到其应该具有的功能和执行过程,从而帮助读者正确实现此函数。

图1-5 工具栏上的"演示模式"按钮

当工具栏上的"演示模式"按钮没有高亮显示时(单击工具栏上的"演示模式"按钮可以使其切换状态),CP Lab处于非演示模式。在非演示模式下调试观察点函数时,会使用其函数体中的代码和断点。

3.10 在演示模式下调试项目

读者可以按照下面的步骤,练习在演示模式下调试项目(主要是调试观察点函数):
(1) 保证工具栏上的"演示模式"按钮高亮显示。
(2) 在"调试"菜单中选择"启动调试"(快捷键是F5)。

启动调试后,程序会在观察点函数的开始位置处中断,如图1-6所示。源代码编辑器左侧空白处显示了相应的图标,分别标识了观察点函数的起始位置和结束位置,以及下一行要执行的代码(黄色箭头)。

图1-6 启动调试后,在观察点函数的开始位置中断

同时,在"转储信息"窗口中(可以选择"调试"菜单"窗口"中的"转储信息"打开此窗口)显示了观察点函数正在操作的数据信息,如图1-7所示。数据信息主要包含如以下内容:
(1) 函数调用信息。对本次观察点函数的调用信息进行了描述。
(2) 函数返回信息。由于此时刚刚进入观察点函数,所以还无法显示其返回信息。当在观察点函数结束位置中断时,即可显示其返回信息。主要对观察点函数的返回值或者操作结果进行描述。
(3) 重要的数据信息。包含了正则表达式和解析树的后序序列的描述,以及对栈中保存的NFA片段的信息进行了描述,包括状态名称(用数字表示从1开始)转换标志以

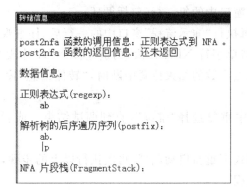

图 1-7 在观察点函数开始位置中断时的"转储信息"窗口

及转换到的状态。当程序运行到观察点结尾的位置时还会显示返回结果用以验证项目（验证项目会在后面详细描述）。

(4) 在"调试"菜单中选择"继续"（快捷键是 F5）。

由于是在"演示模式"下调试观察点函数，CP Lab 会忽略掉函数体中的所有代码，取而代之的是使用 CP Lab 提供的演示功能对观察点函数的执行过程进行演示，所以，CP Lab 会自动打开"演示流程"窗口（可以选择"调试"菜单"窗口"中的"演示流程"打开此窗口），在其中显示观察点函数的演示流程，如图 1-8 所示。

```
State* post2nfa(char* postfix)
{
    char *p;                                // 游标
    Fragment fragment1, fragment2, fm;      // NFA 片段结构体变量
    Fragment fragment = {0, 0};             // 初始化用于返回的 NFA 片段
    State *NewStartState, *NewAcceptState;  // 开始状态和接受状态
    for(p=postfix; *p!=NULL; p++)
    {
        switch(*p)
        {
            default:    // 构造单字符 NFA 片段
                // 调用 CreateState 函数生成两个新的状态
                NewStartState = CreateState();
                NewAcceptState = CreateState();
                // 开始状态通过单字符转换到接受状态
                NewStartState->Transform = *p;
                NewStartState->Next1 = NewAcceptState;
                NewAcceptState->AcceptFlag = 1;
                // 调用 MakeFragment 函数生成一个新的 NFA 片段，并入栈
                fm = MakeFragment(NewStartState, NewAcceptState);
                Push(&FragmentStack, fm);
                break;
            case '.':   // 构造连接 NFA 片段
                // 栈顶的两个片段出栈，构造新的 NFA 片段
                fragment2 = Pop(&FragmentStack);
```

图 1-8 "演示流程"窗口

观察点函数的演示流程通常采用简洁、直观的语言进行描述（一行描述可能会对应多行 C 源代码），偶尔也会在读者理解起来比较困难的地方提供 C 源代码的提示或者直接使用 C 源代码，目的就是为了方便读者将演示流程快速转换为 C 源代码。在"演示流程"窗口左侧的空白处，同样使用黄色箭头标识出了下一行要执行的代码（流程）。

(5) 在"调试"菜单中重复选择"继续"，直到在观察点函数的结束位置中断。CP Lab

实验 1 实验环境的使用

会单步执行"演示流程"窗口中的每一行(包括循环)。

在调试的过程中,每执行"演示流程"窗口中的一行后,仔细观察"转储信息"窗口内容所发生的变化,例如构造单字符 NFA 片段,构造连接 NFA 片段等,理解 NFA 片段构造的执行过程。当在观察点函数的结束位置中断时,"转储信息"窗口中将显示函数的返回信息。

(6) 在"调试"菜单中重复选择"继续",直到调试结束。或者,在"调试"菜单中选择"停止调试"。

读者可以在演示模式下重新启动调试,再次执行以上的步骤,仔细体会在"演示模式"下调试观察点函数的过程。

3.11 验证项目(失败)

这里介绍 CP Lab 提供的另外一个重要功能:验证功能。

之前提到了 main.c 文件中的 post2nfa 函数还不完整,是留给读者完成的。而当读者完成此函数后,往往需要使用调试功能或者执行功能来判断所完成的函数是否能够达到预期的效果,即是否与演示时函数的执行过程和返回值完全一致。CP Lab 提供的验证功能可以自动化地、精确地完成这个验证过程。

验证功能分为下面 3 个阶段:

(1) 在"演示模式"下执行观察点函数(与工具栏上的"演示模式"按钮是否高亮无关),将产生的转储信息自动保存在文本文件 ValidateSource.txt 中。

(2) 在"非演示模式"下执行观察点函数,将产生的转储信息自动保存在文本文件 ValidateTarget.txt 中。

(3) 自动使用 CP Lab 提供的文本文件比较工具来比较这两个文件。当这两个文件中的转储信息完全一致时,报告"验证成功";否则,报告"验证失败"。

当读者完成的函数与演示时函数的执行过程和返回值完全一致时,就会产生完全一致的转储信息,验证功能就会报告"验证成功";否则,验证功能就会报告"验证失败",并且允许读者使用 CP Lab 提供的文本文件比较工具来查看这两个转储信息文件中的不同之处,从而帮助读者迅速、准确地找到验证失败的原因,进而继续修改源代码,直到验证成功。

按照下面的步骤启动验证功能:

(1) 在"调试"菜单中选择"开始验证"(快捷键是 Alt+F5)。在验证过程中,"输出"窗口会实时显示验证各个阶段的执行过程(如清单 1-1 所示),包括转储信息文件的路径、观察点函数的调用信息和返回信息以及验证结果。由于 post2nfa 函数还不完整,所以验证失败。

(2) 使用"输出"窗口工具条上的"比较"按钮(如图 1-9 所示)查看两个转储信息文件中的内容,即它们之间的不同之处。

提示:

- 在转储信息文件中,只保存了观察点函数开始位置和结束位置的转储信息,并使用单线进行分隔。观察点函数多次被调用的转储信息之间使用双线进行分隔。

- 在转储信息文件中,为了确保验证功能的准确性,某些信息会被忽略掉(不再显示或使用"N/A"替代),例如内存中的随机值等。

清单 1-1:在"输出"窗口中显示的验证信息

------已启动验证:项目:RegexpToNFA,配置:Debug------

验证第一阶段:正在使用"演示模式"生成转储信息,并写入源文件…
源文件路径:G:\Documents and Settings\Engintime\My Documents\CP Lab\Projects\lab1\Debug\ValidateSource.txt

post2nfa 函数的调用信息:正则表达式到 NFA。
post2nfa 函数的返回信息:返回 NFA 的开始状态地址。

==

验证第二阶段:正在使用"非演示模式"生成转储信息,并写入目标文件…
目标文件路径:G:\Documents and Settings\Engintime\My Documents\CP Lab\Projects\lab1\Debug\ValidateTarget.txt

post2nfa 函数的调用信息:正则表达式到 NFA。
post2nfa 函数的返回信息:返回 NFA 的开始状态地址。

==

验证第三阶段:正在比较转储信息源文件与目标文件的内容…
比较结果:转储信息源文件与目标文件的内容不同

===================验证结果:失败===================

图 1-9 "输出"窗口工具栏上的"比较"按钮

3.12 实现 post2nfa 函数

文件 main.c 中的 post2nfa 的函数体还不完整,需要读者补充完整。
提示:

- 在"观察点"窗口中,可以在函数名称上右击,选择快捷菜单中的"查看演示流程",CP Lab 会打开"演示流程"窗口,并显示观察点函数的演示流程。这样,即使在没有启动调试的情况下,读者也可以方便地查看观察点函数的演示流程。
- NFA 状态图例如图 1-10 所示。参考演示流程中构造 NFA 片段的描述,其中给出了构造单字符 NFA 片段(图 1-11 所示)和连接 NFA 片段的源代码,这两步操作可以完成对例 1 正则表达式到 NFA 的转换(转换后的 NFA 如图 1-12 所示),

读者可以在"演示模式"下一边调试一边理解源代码的执行过程,在此基础上完成其他形式 NFA 片段的构造。其中,例 2 是正则表达式到 NFA 的转换后的 NFA,如图 1-13 所示;例 3 是正则表达式到 NFA 的转换后的 NFA,如图 1-14 所示;例 5 是正则表达式到 NFA 的转换后的 NFA,如图 1-15 所示。

图 1-10　NFA 状态图例　　　　　　图 1-11　表示单字符的 NFA 片段

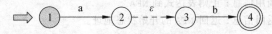

图 1-12　表示连接的 NFA 片段(对应例 1)

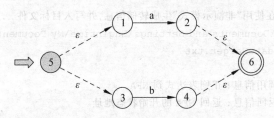

图 1-13　表示选择的 NFA 片段(对应例 2)

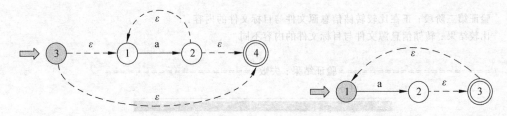

图 1-14　表示星号的 NFA 片段(对应例 3)　　　图 1-15　表示加号的 NFA 片段(对应例 5)

- 对于问号的 NFA 片段的构造(如图 1-16 所示),原则上也可以将状态 1 作为开始状态,状态 2 作为接受状态,并通过 ε 转换来表达接受状态为空的情况。但是如果存在一个或多个 NFA 片段与问号 NFA 片段相连接的情况,上述处理办法就不适用了。因为在本程序中使用的是类似于二叉树的存储结构,一个状态最多只有两个转换指针,所以,必须在原来的基础上再添加两个状态作为开始状态和接受状态。

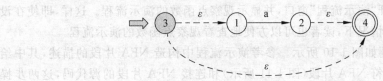

图 1-16　表示问号的 NFA 片段(对应例 4)

3.13 在非演示模式下调试项目

读者在实现了 post2nfa 函数后,可以按照下面的步骤,练习在非演示模式下调试项目(主要是调试由读者实现的观察点函数):

(1) 在"生成"菜单中选择"生成项目"。如果读者编写的源代码中存在语法错误,修改这些错误,直到可以成功生成项目。

(2) 单击工具栏上的"演示模式"按钮,使其切换到非高亮显示状态。

(3) 在"调试"菜单中选择"启动调试"。程序会在观察点函数的开始位置处中断。

(4) 在"调试"菜单中重复选择"逐过程"(快捷键是 F10),直到在观察点函数的结束位置中断。CP Lab 会单步执行观察点函数中的每一行源代码。在调试的过程中,每执行一行源代码后,仔细观察"转储信息"窗口内容所发生的变化,例如游标的移动、构造单字符的 NFA 片段等,理解 NFA 片段的构造过程。当在观察点函数的结束位置中断时,"转储信息"窗口中将显示函数的返回信息。

以上的练习说明,CP Lab 可以让读者在非演示模式下调试项目,并观察"转储信息"窗口内容所发生的变化,从而理解每一行源代码对内存数据的操作结果。如果读者发现所编写的源代码存在异常行为(例如死循环、数组越界访问或者验证失败),可以在非演示模式下单步调试项目,来查找异常产生的原因。

3.14 验证项目(成功)

按照下面的步骤启动验证功能:
(1) 在"调试"菜单中选择"开始验证"。
(2) 如果验证失败,读者可以参考之前的内容来查找原因并修改源代码中的错误,直到验证成功。

3.15 提交作业

如果读者写完了程序并通过了自动化验证,可以使用"提交作业"功能,将读者编写的源代码文件自动提交到服务器,供教师查看。步骤如下:

(1) 选择"用户"菜单中的"提交作业",打开"提交作业"对话框。
(2) 在"提交作业"对话框中,单击"继续提交"按钮,完成提交作业操作。
(3) 如果需要重新绑定课时,单击"重新绑定课时"按钮,绑定符合要求的课时。

注意:可以多次执行"提交作业"操作,后提交的作业会覆盖之前提交的作业。在未登录的情况下,不能提交作业。

3.16 总结

读者使用 CP Lab 进行编译原理实验的步骤可以总结如下:
(1) 启动 CP Lab。
(2) 注册新用户,或使用已有用户登录。
(3) 新建实验项目,并绑定课时。

(4) 在演示模式下调试项目,理解观察点函数的执行过程(通常观察点函数还未完整实现)。

(5) 结合观察点函数的演示流程,修改观察点函数的源代码,实现其功能。

(6) 生成项目(排除所有的语法错误)。

(7) 验证观察点函数。如果验证失败,可以使用"比较"功能,或者在非演示模式下调试项目,从而定位错误的位置,然后回到步骤(5)。

(8) 提交作业

(9) 退出 CP Lab。

3.17 获得帮助

如果读者在使用 CP Lab 的过程中遇到问题需要专业的解答,或者有一些心得体会想和其他 CP Lab 用户分享,欢迎加入 CP Lab 网上论坛。加入论坛的方法有两种:

- 选择 CP Lab"帮助"菜单中的"论坛"。
- 直接访问 http://www.engintime.com/forum。

下面列出了读者在使用 CP Lab 的过程中可能遇到的一些问题和使用技巧,用于帮助读者更好地使用 CP Lab,获得最佳的实验效果。

(1) 读者时常会遇到在自己编写的源代码中存在死循环的情况,这就会造成 CP Lab 的调试功能,特别是验证功能无法自行结束。此时,读者可以选择"调试"菜单中的"停止调试"(快捷键是 Shift+F5)来强制结束这些功能。随后,读者可以检查自己编写的源代码,或者在"非演示模式"下单步调试项目,从而找到造成死循环的原因。

(2) 读者时常会遇到的另外一个情况是"数组越界访问"。此时,CP Lab 会弹出一个调试异常对话框,读者只要选择对话框中的"是"按钮,就可以立即定位到异常所在的代码行。

(3) CP Lab 作为一个 IDE 环境,提供了强大的调试功能,包括单步调试、添加断点、查看变量的值、查看调用堆栈等。读者在调试过程中可以灵活使用这些功能,提高调试效率。注意,在"演示模式"下,观察点函数中的断点会被忽略。

(4) 在"演示模式"下,对观察点函数只能进行单步调试(无论是按快捷键 F5,还是 F10),如果观察点函数中存在多次循环,会造成调试过程比较缓慢。此时,读者可以选择"调试"菜单中的"结束观察"(快捷键是 Shift+Alt+F5),直接跳转到观察点函数的结束位置中断。

(5) "输出"窗口、"演示流程"窗口以及"转储信息"窗口中的文本信息可以被选中并复制(但是不能修改),读者可以很方便地将这些信息保存下来,用于完成实验报告等工作。

(6) CP Lab 提供的实验项目通常不会在 Windows 控制台窗口中打印输出任何信息,因为在"转储信息"窗口、"输出"窗口中已经为读者提供了足够多的信息。当然,读者也可以根据自己的喜好,使用 printf 函数,在 Windows 控制台窗口中打印输出一些信息(这些信息不会对"验证"结果产生任何影响)。如果读者想快速查看程序在 Windows 控制台窗口中打印输出的信息,可以使用"调试"菜单中的"开始执行"功能(快捷键是 Ctrl+

F5 键)。

(7) CP Lab 为读者提供了查看用户信息、修改用户信息、修改密码、修改密保问题和找回密码等功能。

四、思考与练习

1. 编写一个 FreeNFA 函数,当在 main 函数的最后调用此函数时,可以将整个 NFA 的内存释放掉,从而避免内存泄露。

2. 读者编写完代码之后可以对 main.c 中 main 函数之前的例 2 到例 5 进行一一验证,确保程序可以将所有形式的正则表达式转换为正确的 NFA,并验证通过。

3. 对例 6、例 7、例 8 的正则表达式进行验证,并画出例 7 和例 8 的 NFA 状态图。

4. 详细阅读 re2post 函数中的源代码,并尝试在源代码中添加注释。然后尝试为本实验中所有的例子绘制解析树(类似二叉树)。

EXPERIMENT 2

实验 2　NFA 到 DFA

实验难度：★★★★☆
建议学时：4 学时

一、实验目的

- 掌握 NFA 和 DFA 的概念。
- 掌握 ε-闭包的求法和子集的构造方法。
- 实现 NFA 到 DFA 的转换。

二、预备知识

- 完成从正则表达式到 NFA 的转换过程是完成本实验的先决条件。虽然 DFA 和 NFA 都是典型的有向图，但是基于 NFA 自身的特点，在之前使用了类似二叉树的数据结构来存储 NFA，达到了简化的目的。但是，DFA 的结构相对复杂，所以在这个实验中使用了图的邻接链表来表示 DFA。如果读者对有向图的概念和邻接链表表示法有一些遗忘，可以复习一下数据结构中相关的章节。
- 对 DFA 的含义有初步的理解，了解 ε-闭包的求法和子集的构造方法。读者可以参考配套的《编译原理》教材，预习这一部分内容。

三、实验内容

3.1　准备实验

按照下面的步骤准备实验：
（1）启动 CP Lab。
（2）在"文件"菜单中选择"新建"|"项目"，打开"新建项目"对话框。
（3）使用模板"002 NFA 到 DFA"新建一个项目。

3.2　阅读实验源代码

1. NFAToDFA.h 文件（参见源代码清单 2-1）

此文件主要定义了与 NFA 和 DFA 相关的数据结构，其中有关 NFA 的数据结构在前一个实验中有详细说明，所以这里主要说明一下有关 DFA 的 3 个数据结构。这些数据结构定义了 DFA 的邻接链表，其中，DFAState 结构体用于定义有向图中的顶点（即 DFA 状态），Transform 结构体用于定义有向图中的弧（即转换）。具体内容可参见表 2-1 和表 2-2。

表 2-1

DFA 的域	说　明
DFAlist	DFA 状态集合
length	集合中的 DFA 状态数量。与前一项构成一个线性表

表 2-2

DFAState 的域	说　明
NFAlist	NFA 状态集合。用于保存 DFA 状态中的 NFA 状态集合
NFAStateCount	集合中的 NFA 状态数量。与前一项构成一个线性表
firstTran	指向第一个转换

Transform 的域	说　明
TransformChar	状态之间的转换符号
DFAStateIndex	DFA 状态在线性表中的下标。用于指示转换的目标 DFA 状态
NFAlist	NFA 状态集合。用于保存构造的子集以及生成新的 DFA 状态
NFAStateCount	集合中的 NFA 状态数量。与前一项构成一个线性表
NextTrans	指向下一个转换

2．main.c 文件(参见源代码清单 2-2)

此文件定义了 main 函数。在 main 函数中首先初始化栈，然后调用 re2post 函数，将正则表达式转换到解析树的后序序列，最后调用 post2dfa 函数将解析树的后序序列转换到 DFA。

在 main 函数的后面定义了一系列函数，有关函数的具体内容参见表 2-3。关于这些函数的参数和返回值，可以参见其注释。

表 2-3

函　数　名	功　能　说　明
CreateDFATransform	创建一个新的 DFA 转换。每当求得一个 ε-闭包后可以调用此函数来创建一个 DFA 转换
CreateDFAState	利用转换作为参数构造一个新的 DFA 状态，同时将 NFA 状态子集复制到新创建的 DFA 状态中
NFAStateIsSubset	当一个子集构造完成后，需要调用此函数来判断是否需要为该子集创建一个新的 DFA 状态。如果构造的子集是某一个 DFA 状态中 NFA 状态集合的子集，就不需要新建 DFA 状态了。此函数的函数体还不完整，留给读者完成
IsTransformExist	判断 DFA 状态的转换链表中是否已经存在一个字符的转换。每当求得一个 ε-闭包后可以调用此函数来决定是新建一个转换，还是将 ε-闭包合并到已有的转换中。此函数的函数体还不完整，留给读者完成
AddNFAStateArrayToTransform	将一个 NFA 集合合并到一个 DFA 转换的 NFA 集合中，并确保重复的 NFA 状态只出现一次。当需要将 ε-闭包合并到已有的转换中的 NFA 集合中时，可以调用此函数。此函数的函数体还不完整，留给读者完成

续表

函　数　名	功能说明
Closure	使用二叉树的先序遍历算法求一个 NFA 状态的 ε-闭包。注意,优先使用二叉树的先序遍历算法,否则会造成 ε-闭包中 NFA 状态集合顺序不同,进而导致无法通过自动化验证。此函数的函数体还不完整,留给读者完成
post2dfa	将解析树的后序序列转换为 DFA。在这个函数中会调用函数 post2nfa,所以在此函数中只需要专注于 NFA 转换到 DFA 的源代码的编写即可。此函数的函数体还不完整,留给读者完成

3. RegexpToPost.c 文件(参见源代码清单 2-3)

此文件定义了 re2post 函数,此函数的主要功能是将正则表达式转换为解析树的后序序列形式。

4. PostToNFA.c 文件(参见源代码清单 2-4)

此文件定义了 post2nfa 函数,此函数的主要功能是将解析树的后序序列形式转换为 NFA。关于此函数的功能、参数和返回值,可以参见其注释。注意,此函数的函数体还不完整,读者可以直接使用之前实验中编写的代码。

5. NFAFragmentStack.c 文件(参见源代码清单 2-5)

此文件定义了与栈相关的操作函数。注意,这个栈是用来保存 NFA 片段的。

6. NFAStateStack.c 文件(参见源代码清单 2-6)

此文件定义了与栈相关的操作函数。注意,这个栈是用来保存 NFA 状态的。

7. RegexpToPost.h 文件(参见源代码清单 2-7)

此文件声明了相关的操作函数。为了使程序模块化,将 re2post 函数声明包含在一个头文件中,再将此头文件包含到 main.c 中。

8. PostToNFA.h 文件(参见源代码清单 2-8)

此文件声明了相关的操作函数。为了使程序模块化,将 post2nfa 函数声明包含在一个头文件中,再将此头文件包含到 main.c 中。

9. NFAFragmentStack.h 文件(参见源代码清单 2-9)

此文件定义了与栈相关的数据结构并声明了相关的操作函数。

10. NFAStateStack.h 文件(参见源代码清单 2-10)

此文件定义了与栈相关的数据结构并声明了相关的操作函数。

源代码清单 2-1: NFAToDFA.h 文件

```
#ifndef _NFATODFA_H_
#define _NFATODFA_H_

//
//在此处包含 C 标准库头文件
//
```

```c
#include<stdio.h>

//
//在此处包含其他头文件
//

//
//在此处定义数据结构
//

#define MAX_STATE_NUM 64                    //状态的最大数量

typedef struct _NFAState{
    char Transform;                 //状态间转换的标识,用'$'表示'ε-转换'
    struct _NFAState * Next1;       //指向下一个状态
    struct _NFAState * Next2;       //指向下一个状态
    int Name;                       //状态名称
    int AcceptFlag;    //是否为接受状态的标志,1表示是接受状态,0表示不是接受状态
}NFAState;

//Fragment 结构是一个 NFA 的片段
typedef struct _NFAFragment{
    NFAState * StartState;          //开始状态
    NFAState * AcceptState;         //接受状态
}NFAFragment;

//转换
typedef struct _Transform{
    char TransformChar;                     //状态之间的转换标识符
    int DFAStateIndex;                      //DFA 状态在数组中的下标
    NFAState * NFAlist[MAX_STATE_NUM];      //NFA 状态集合
    int NFAStateCount;                      //状态集合计数
    struct _Transform * NextTrans;          //指向下一个转换
}Transform;

                                            //DFA 状态
typedef struct _DFAState{
    NFAState * NFAlist[MAX_STATE_NUM];      //NFA 状态集合
    int NFAStateCount;                      //状态集合计数
    Transform * firstTran;                  //指向第一个转换
}DFAState;

//DFA
```

```c
typedef struct DFA{
    DFAState * DFAlist[MAX_STATE_NUM];              //DFA 状态集合
    int length;                                      //状态集合计数
}DFA;

//
//在此处声明函数
//

void Closure(NFAState * State, NFAState * * StateArray, int * Count);
DFAState * CreateDFAState(Transform * pTransform);
Transform * CreateDFATransform (char TransformChar, NFAState * * NFAStateArray,
                                int Count);
int NFAStateIsSubset(DFA * pDFA, Transform * pTransform);
Transform * IsTransformExist(DFAState * pDFAState, char TransformChar);
void AddNFAStateArrayToTransform(NFAState * * NFAStateArray, int Count,
                                Transform * pTransform);
DFA * post2dfa(DFA * pDFA, char * postfix);

//
//在此处声明全局变量
//

extern const char VoidTrans;
extern NFAState * Start;

#endif /* _NFATODFA_H_ */
```

源代码清单 2-2: main.c 文件

```c
#include "NFATODFA.h"
#include "RegexpToPost.h"
#include "PostToNFA.h"
#include "NFAStateStack.h"
#include "NFAFragmentStack.h"

NFAFragmentStack FragmentStack;     //栈,用于存储 NFA 片段
NFAStateStack StateStack;           //栈,用于存储 NFA 状态

const char VoidTrans='$';           //表示空转换

char * regexp="a(a|1) * ";          //例 1
                                    //char * regexp="(aa|b) * a(a|bb) * ";  //例 2
                                    //char * regexp="(a|b) * a(a|b)?";       //例 3
```

```
int main(int argc, char * * argv)
{
    char * post;
    DFA * dfa=(DFA * )malloc(sizeof(DFA));
    dfa->length=0;

    //
    //初始化栈
    //
    InitNFAFragmentStack(&FragmentStack);

    //
    //调用 re2post 函数将正则表达式字符串转换为解析树的后序遍历序列
    //
    post=re2post(regexp);

    //
    //调用 post2dfa 函数将解析树的后序遍历序列转换为 DFA
    //
    dfa=post2dfa(dfa, post);

    return 0;
}

/*  .
功能:
    创建一个 DFA 状态的转换。

参数:
    TransformChar--转换符号。
    NFAStateArray--NFA 状态指针数组。
    Count--数组元素个数。

返回值:
     Transform 结构体指针。
*/
Transform * CreateDFATransform (char TransformChar, NFAState * *NFAStateArray,
                        int Count)
{
    int i;
    Transform * pTransform= (Transform * )malloc(sizeof(Transform));

    for(i=0; i<Count; i++)
```

实验 2 NFA 到 DFA

```
            pTransform->NFAlist[i]=NFAStateArray[i];

    pTransform->NFAStateCount=Count;
    pTransform->TransformChar=TransformChar;
    pTransform->DFAStateIndex=-1;
    pTransform->NextTrans=NULL;

    return pTransform;
}
```

/*
功能:
    创建一个 DFA 状态。

参数:
    pTransform--DFA 状态转换指针。

返回值:
    DFAState 结构体指针。
*/
```
DFAState * CreateDFAState(Transform * pTransform)
{
    int i;
    DFAState * pDFAState=(DFAState * )malloc(sizeof(DFAState));

    for(i=0; i<pTransform->NFAStateCount; i++)
        pDFAState->NFAlist[i]=pTransform->NFAlist[i];

    pDFAState->NFAStateCount=pTransform->NFAStateCount;
    pDFAState->firstTran=NULL;

    return pDFAState;
}
```

/*
功能:
    判断一个转换中的 NFA 状态集合是否为某一个 DFA 状态中 NFA 状态集合的子集。

参数:
    pDFA--DFA 指针。
    pTransform--DFA 状态转换指针。

返回值:

        如果存在,返回 DFA 状态下标;不存在,返回-1。
*/
int NFAStateIsSubset(DFA * pDFA, Transform * pTransform)
{
    //
    //TODO:在此处添加代码
    //
}

/*
功能:
    判断某个 DFA 状态的转换链表中是否已经存在一个字符的转换。

参数:
    pDFAState--DFAState 指针。
    TransformChar--转换标识符。

返回值:
    Transform 结构体指针。
*/
Transform * IsTransformExist(DFAState * pDFAState, char TransformChar)
{
    //
    //TODO:在此处添加代码
    //
}

/*
功能:
    将一个 NFA 集合合并到一个 DFA 转换中的 NFA 集合中。
    注意,合并后的 NFA 集合中不应有重复的 NFA 状态。

参数:
    NFAStateArray--NFA 状态指针数组,即待加入的 NFA 集合。
    Count--待加入的 NFA 集合中的元素个数。
    pTransform--转换指针。
*/
void AddNFAStateArrayToTransform(NFAState * * NFAStateArray, int Count, Transform * pTransform)
{
    //
    //TODO:在此处添加代码
    //

实验 2 NFA 到 DFA

```
}

/*
功能：
    使用二叉树的先序遍历算法求一个 NFA 状态的 ε-闭包。

参数：
    State--NFA 状态指针。从此 NFA 状态开始求 ε-闭包。
    StateArray--NFA 状态指针数组。用于返回 ε-闭包。
    Count--元素个数。     用于返回 ε-闭包中 NFA 状态的个数。
*/
void Closure(NFAState * State, NFAState * * StateArray, int * Count)
{
    InitNFAStateStack(&StateStack); //调用 InitNFAStateStack 函数初始化栈

    //
    //TODO:在此添加代码
    //
}

/*
功能：
    将解析树的后序遍历序列转换为 DFA。

参数：
    pDFA--DFA 指针。
    postfix--正则表达式的解析树后序遍历序列。

返回值：
    DFA 指针。
*/
NFAState * Start=NULL;
DFA * post2dfa(DFA * pDFA, char * postfix)
{
    int i, j;                                        //游标
    Transform * pTFCursor;                           //转换指针
    NFAState * NFAStateArray[MAX_STATE_NUM];         //NFA 状态指针数组,用于保存 ε-闭包
    int Count=0;                                     //ε-闭包中的元素个数

    //
    //调用 post2nfa 函数将解析树的后序遍历序列转换为 NFA,并返回开始状态
    //
    Start=post2nfa(postfix);
```

```
    //
    //TODO:在此添加代码
    //

    return pDFA;
}
```

源代码清单 2-3: RegexpToPost.c 文件

```
#include "RegexpToPost.h"

/*
功能:
    将输入的正则表达式字符串转换为解析树的后续遍历序列。

参数:
    re--正则表达式指针。

返回值:
    解析树的后续遍历序列数组指针。
*/
char * re2post(char * re)
{
    int nalt;      //表示解析到这个字符为止,已经有多少个分支结构
    int natom;     //表示解析到这个字符为止,已经有多少个原子结构
    static char buf[8000];
    char * dst;

    struct {
        int nalt;
        int natom;
    } paren[100], * p;

    p=paren;
    dst=buf;
    nalt=0;
    natom=0;
    if(strlen(re)>=sizeof buf/2)
        return NULL;

    for(; * re; re++){
        switch(* re){
```

```
            case '(':
                if(natom>1)
                {
                    --natom;
                    *dst++='.';
                }
                if(p>=paren+100)
                    return NULL;
                p->nalt=nalt;
                p->natom=natom;
                p++;
                nalt=0;
                natom=0;
                break;
            case '|':
                if(natom==0)
                    return NULL;
                while(--natom>0)
                    *dst++='.';
                nalt++;
                break;
            case ')':
                if(p==paren)
                    return NULL;
                if(natom==0)
                    return NULL;
                while(--natom>0)
                    *dst++='.';
                for(; nalt>0; nalt--)
                    *dst++='|';
                --p;
                nalt=p->nalt;
                natom=p->natom;
                natom++;
                break;
            case '*':
            case '+':
            case '?':
                if(natom==0)
                    return NULL;
                *dst++=*re;
                break;
            default:
```

```
                    if(natom>1)
                    {
                        --natom;
                        *dst++='.';
                    }
                    *dst++=*re;
                    natom++;
                    break;
            }
        }
        if(p!=paren)
            return NULL;
        while(--natom>0)
            *dst++='.';
        for(;nalt>0;nalt--)
            *dst++='|';
        *dst=0;

        return buf;
    }
```

源代码清单 2-4: PostToNFA.c 文件

```
#include "PostToNFA.h"
#include "NFAFragmentStack.h"

NFAFragmentStack FragmentStack;

/*
功能：
    初始化一个 NFA 状态。

返回值：
    状态结构体指针。
*/
int nstate=1;
NFAState * CreateNFAState()
{
    NFAState * s=(NFAState *)malloc(sizeof(NFAState));

    s->Name=nstate++;
    s->Transform='\0';
    s->Next1=NULL;
    s->Next2=NULL;
```

```
    s->AcceptFlag=0;

    return s;
}

/*
功能：
    初始化 NFAFragment 结构体。

参数：
    StartState--开始状态。
    AcceptState--接受状态。

返回值：
    NFAFragment 结构体指针。
*/
NFAFragment MakeNFAFragment(NFAState * StartState, NFAState * AcceptState)
{
    NFAFragment n={StartState, AcceptState};
    return n;
}

/*
功能：
    将解析树的后序序列转换为 NFA。

参数：
    postfix--解析树的后序序列指针。

返回值：
    NFA 的开始状态指针。
*/
NFAState * post2nfa(char * postfix)
{
    char * p;                                    //游标
    NFAFragment fragment1, fragment2, fm;        //NFA 片段结构体变量
    NFAFragment fragment={0, 0};                 //初始化用于返回的 NFA 片段
    NFAState * NewStartState, * NewAcceptState;  //开始状态和接受状态指针

    //
    //TODO: 在此添加代码
    //

    return fragment.StartState;
}
```

源代码清单 2-5：**NFAFragmentStack.c** 文件

```c
#include "NFAFragmentStack.h"

/*
功能：
    初始化栈。

参数：
    pS--栈的指针。
*/
void InitNFAFragmentStack(NFAFragmentStack * pS)
{
    pS->top=-1;
}

/*
功能：
    将元素入栈。

参数：
    pS--栈的指针。
    Elem--入栈的元素。

返回值：
    空。
*/
void PushNFAFragment(NFAFragmentStack * pS, NFAFragment Elem)
{
    //
    //栈满,入栈失败
    //
    if(MAX_STACK_LENGTH-1<=pS->top)
        return;

    pS->top++;
    pS->buffer[pS->top]=Elem; //将元素插入栈顶

    return;
}

/*
功能：
    将栈顶元素出栈。
```

参数:
    pS--栈的指针。

返回值:
    如果出栈成功,返回出栈元素的值。
    如果出栈失败,返回-1。
*/
NFAFragment PopNFAFragment(NFAFragmentStack * pS)
{
    int pos;
    NFAFragment fragment={0,0};

    //
    //栈为空,出栈失败
    //
    if(NFAFragmentStackEmpty(pS))
        return fragment;

    pos=pS->top;
    pS->top--;

    return pS->buffer[pos];
}

/*
功能:
    判断栈是否为空。

参数:
    pQ--栈的指针。

返回值:
    如果栈空,返回1(真)。
    如果栈非空,返回0(假)。
*/
int NFAFragmentStackEmpty(NFAFragmentStack * pS)
{
    return-1==pS->top ? 1 : 0;
}

源代码清单 2-6: **NFAStateStack.c** 文件

```
#include "NFAToDFA.h"
#include "NFAStateStack.h"
```

```
/*
功能：
    初始化栈。

参数：
    pS--栈的指针。
*/
void InitNFAStateStack(NFAStateStack * pS)
{
    pS->top=-1;
}


/*
功能：
    将元素入栈。

参数：
    pS--栈的指针。
    Elem--入栈的元素。

返回值：
    空。
*/
void PushNFAState(NFAStateStack * pS, NFAState * Elem)
{
    //
    //栈满,入栈失败
    //
    if(MAX_STACK_LENGTH-1<=pS->top)
        return;

    pS->top++;
    pS->buffer[pS->top]=Elem;          //将元素插入栈顶

    return;
}


/*
功能：
    将栈顶元素出栈。

参数：
    pS--栈的指针。
```

返回值：
    如果出栈成功,返回出栈元素的值。
    如果出栈失败,返回-1。
*/
```c
NFAState * PopNFAState(NFAStateStack * pS)
{
    int pos;
    NFAState * State=0;

    //
    //栈为空,出栈失败
    //
    if(NFAStateStackEmpty(pS))
        return State;

    pos=pS->top;
    pS->top--;

    return pS->buffer[pos];
}
```

/*
功能：
    判断栈是否为空。

参数：
    pQ--栈的指针。

返回值：
    如果栈空,返回1(真)。
    如果栈非空,返回0(假)。
*/
```c
int NFAStateStackEmpty(NFAStateStack * pS)
{
    return-1==pS->top ? 1 : 0;
}
```

源代码清单 2-7：RegexpToPost.h 文件

```c
#ifndef _REGEXPTOPOST_H_
#define _REGEXPTOPOST_H_

//
//在此处包含C标准库头文件
//
```

```c
////在此处包含其他头文件
//

#include "NFAToDFA.h"

//
//在此处定义数据结构
//

//
//在此处声明函数
//

char * re2post(char * re);

#endif /* _REGEXPTOPOST_H_ */
```

源代码清单 2-8: PostToNFA.h 文件

```c
#ifndef _POSTTONFA_H_
#define _POSTTONFA_H_

//
//在此处包含 C 标准库头文件
//

//
//在此处包含其他头文件
//

#include "NFAToDFA.h"

//
//在此处定义数据结构
//

//
//在此处声明函数
//

NFAState * CreateNFAState();
NFAFragment MakeNFAFragment(NFAState * StartState, NFAState * AcceptState);
```

```
NFAState * post2nfa(char * postfix);

//
//在此处声明全局变量
//

#endif /* _POSTTONFA_H_ */
```

源代码清单 2-9：NFAFragmentStack.h 文件

```
#ifndef _NFAFRAGMENTSTACK_H_
#define _NFAFRAGMENTSTACK_H_

//
//在此处包含 C 标准库头文件
//


//
//在此处包含其他头文件
//

#include "NFAToDFA.h"

//
//在此处定义数据结构
//

#define MAX_STACK_LENGTH 1024       //栈的最大长度

//栈
typedef struct _NFAFragmentStack{
    NFAFragment buffer[MAX_STACK_LENGTH];    //栈的缓冲区
    int top;              //指示栈顶的位置,而不是栈中元素的个数
}NFAFragmentStack;

//
//在此处声明函数
//

void InitNFAFragmentStack(NFAFragmentStack * pS);
void PushNFAFragment(NFAFragmentStack * pS, NFAFragment Elem);
NFAFragment PopNFAFragment(NFAFragmentStack * pS);
int NFAFragmentStackEmpty(NFAFragmentStack * pS);
```

```
//
//在此处声明全局变量
//

#endif /* _NFAFRAGMENTSTACK_H_ */
```

源代码清单 2-10：NFAStateStack.h 文件

```
#ifndef _NFASTATESTACK_H_
#define _NFASTATESTACK_H_

//
//在此处包含 C 标准库头文件
//


//
//在此处包含其他头文件
//


//
//在此处定义数据结构
//

#define MAX_STACK_LENGTH 1024        //栈的最大长度

//栈
typedef struct _NFAStateStack{
    NFAState * buffer[MAX_STACK_LENGTH];       //栈的缓冲区
    int top;            //指示栈顶的位置,而不是栈中元素的个数
}NFAStateStack;

//
//在此处声明函数
//

void InitNFAStateStack(NFAStateStack * pS);
void PushNFAState(NFAStateStack * pS, NFAState * Elem);
NFAState * PopNFAState(NFAStateStack * pS);
int NFAStateStackEmpty(NFAStateStack * pS);

//
//在此处声明全局变量
//

#endif /* _NFASTATESTACK_H_ */
```

实验2 NFA 到 DFA

3.3 在演示模式下调试项目

按照下面的步骤调试项目：
(1) 按 F7 键生成项目。
(2) 在演示模式下，按 F5 键启动调试项目。程序会在观察点函数的开始位置中断。
(3) 重复按 F5 键，直到调试过程结束。

在调试的过程中，每执行"演示流程"窗口中的一行后，仔细观察"转储信息"窗口内容所发生的变化，理解 NFA 到 DFA 的转换过程。正则表达式 a(a|1)* 对应的 NFA 状态图如图 2-1 所示，DFA 状态图图 2-2 所示。"转储信息"窗口显示的数据信息包括：

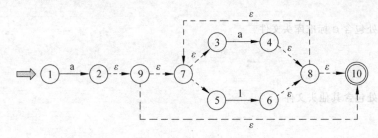

图 2-1 调用 post2nfa 函数后返回的 NFA

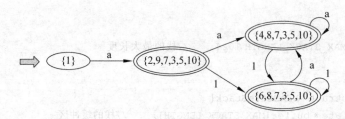

图 2-2 NFA 对应的 DFA

- 正则表达式。
- 解析树的后序序列。
- NFA 向 DFA 转换过程中构造的 ε-闭包。包括元素个数和闭包中的元素。
- DFA 邻接表。包括 DFA 状态在线性表中的下标，DFA 状态中的 NFA 状态集合以及 DFA 状态的转换链表。在调试的最后，DFA 的接受状态会包含在中括号中。
- 调用 post2nfa 函数返回的 NFA。详细的内容可以参看之前实验中的相关说明。

3.4 编写源代码并通过验证

按照下面的步骤继续实验：
(1) 为 post2dfa 函数和其他未完成的函数编写源代码。注意尽量使用已定义的局部变量。
(2) 按 F7 生成项目。如果生成失败，根据"输出"窗口中的提示信息修改源代码中的语法错误。

(3) 按 Alt+F5 键启动验证。如果验证失败,可以使用"输出"窗口中的"比较"功能,或者在"非演示模式"下按 F5 键启动调试后重复按 F10 键单步调试读者编写的源代码,从而定位错误的位置,然后回到步骤(1)。

注意:在实现 Closure 函数时,尽量使用二叉树的先序遍历算法,否则会造成 ε-闭包中 NFA 状态集合顺序不同,进而导致无法通过自动化验证。

四、思考与练习

1. 编写一个 FreeNFA 函数和一个 FreeDFA 函数,当在 main 函数的最后调用这两个函数时,可以将整个 NFA 和 DFA 的内存分别释放掉,从而避免内存泄露。

2. 读者可以尝试使用自己编写的代码将 main 函数之前的例 2 和例 3 转换为 DFA,并确保能够通过自动化验证。在验证通过后,根据 DFA 的邻接链表数据绘制出 DFA 的状态转换图,并尝试编写一个 Minimize 函数,此函数可以将 DFA 中的状态数最小化。

3. 编写一个 Match 函数,此函数可以将一个字符串与正则表达式转换的 DFA 进行匹配,如果匹配成功返回 1,否则返回 0。

EXPERIMENT 3

实验 3　使用 Lex 自动生成扫描程序

实验难度：★★☆☆☆
建议学时：2 学时

一、实验目的

- 掌握 Lex 输入文件的格式。
- 掌握使用 Lex 自动生成扫描程序的方法。

二、预备知识

- 要求已经学习了正则表达式的编写方法，能够正确使用"＊""？""＋"等基本的元字符，并且学习了 Lex 程序中定义的特有的元字符，例如"[ ]""\n"等。
- 了解了标识符和关键字的识别方法。
- 本实验使用 Lex 的一个实现版本——GNU Flex 作为扫描程序。

三、实验内容

3.1 准备实验

按照下面的步骤准备实验：
(1) 启动 CP Lab。
(2) 在"文件"菜单中选择"新建"|"项目"，打开"新建项目"对话框。
(3) 使用模板"003 使用 Lex 自动生成扫描程序"新建一个项目。

3.2 阅读实验源代码

1. sample.txt 文件（参见源代码清单 3-1）

在 sample.txt 文件中是一个使用 TINY 语言编写的小程序，这个小程序在执行时从标准输入（键盘）读取一个整数，计算其阶乘后显示到标准输出（显示器）。

在本实验中并不需要执行这个小程序，只需使用 Lex 生成的扫描程序对这个 TINY 源代码文件进行扫描，最后统计出各种符号的数量。TINY 语言中的符号说明可以参考表 3-1。

表 3-1

序号	符号或语句	说　明
1	{…}	在两个大括号之间的是注释
2	read	从标准输入(键盘)读取数据。是一个关键字
3	write	将数据写入标准输出(屏幕)。是一个关键字
4	if…then…else…end	if 语句。if 的后面是一个布尔表达式,then 和 else 的后面是一个语句块,其中 else 是可选的。end 表示结束。包括了 4 个关键字
5	repeat…until…	repeat 语句。repeat 后面是一个语句块,until 后面是一个布尔表达式。包括了两个关键字
6	<、=	比较运算符。<是小于符号；=是等于符号
7	+、-、*、/	算术运算符
8	:=	赋值运算符
9	正整数	由数字 0~9 组成
10	标识符	由大写字母和小写字母组成
11	;	在语句的结束位置有一个分号

2．define.h 文件(参见源代码清单 3-2)

在此文件中定义了一个枚举类型,对应于 TINY 语言中的各种符号。注意,此文件中还包括了"文件结束"和"错误",其中 yylex 函数在遇到文件结束时,默认会返回 0,所以将"文件结束"定义在开始位置,这样它的值也为 0。

3．scan.txt 文件(参见源代码清单 3-3)

此文件是 Lex 的输入文件。根据 Lex 输入文件的格式,此文件分为 3 个部分(由％％分隔),各个部分的说明可以参见表 3-2。

表 3-2

名　称	说　明
第一部分	• 在％{和％}之间直接插入 C 源代码文件的内容。包括了要包含的头文件,以及 TINY 语言中各种符号的计数器,用于保存扫描的结果 • 定义了换行(newline)和空白(whitespace)的正则表达式。注意,空白包括了空格和制表符
第二部分	包含了一组规则,当规则中的正则表达式匹配时,yylex 函数会执行规则提供的源代码。主要包含下面的规则： • 比较运算符、算术运算符等的规则。这些规则直接使用字符串进行匹配。若匹配成功,yylex 函数返回对应的枚举值 • 匹配换行的规则。匹配成功,就统计行数。 • 匹配空白的规则。匹配成功,就忽略。 • 由于编写注释的正则表达式比较复杂,所以在匹配注释的开始符号后,直接编写 C 代码来匹配注释的结束符号。这里用到的 input 函数是一个 Lex 的内部函数,它会返回一个输入字符 • 使用 Lex 正则表达式中的"."元字符匹配其他的任何字符。当之前的规则均匹配失败时,就会在此规则匹配成功,从而返回错误类型。
第三部分	这部分中的内容会直接插入 C 源代码文件。此部分内容的说明可以参见表 3-3。

表 3-3

内 容	说 明
main 函数	主函数。这里用到了 C 语言定义的 main 函数的两个参数,这两个参数的用法可以参考函数的注释 在 main 函数中首先使用 fopen 函数打开了待处理的文件,然后将此文件作为 Lex 扫描程序的输入(yyin),之后调用 yylex 函数开始扫描,并调用 stat 函数统计各种符号的数量,最后调用 output 函数输出统计的结果
id2keyword 函数	此函数将标识符转换为对应的关键字类型,如果通过参数传入的字符串不能与任何关键字匹配,就仍然返回标识符类型。此函数的函数体还不完整,留给读者完成
stat 函数	此函数根据 tt 参数传入的符号类型,增加符号类型对应的计数器。如果 tt 是标识符类型(ID),会首先调用 id2keyword 函数,尝试将标识符类型转换为对应的关键字类型
output 函数	输出统计的结果

4. main.c 文件

此文件默认是一个空文件。Lex 根据输入文件生成的 C 源代码会输出到此文件中。此文件会包含自动生成的 yylex 函数的定义,此函数实现了与输入文件相对应的 DFA 表驱动。

源代码清单 3-1: **sample.txt** 文件

```
{ Sample program
  int TINY language-
  computes factorial
}
read x; { input an integer }
if 0<x then { don't compute if x<=0 }
  fact :=1;
  repeat
    fact :=fact * x;
    x :=x-1
  until
    x=0;
  write fact { output factorial of x }
end
```

源代码清单 3-2: **define.h** 文件

```
#ifndef _DEFINE_H_
#define _DEFINE_H_

typedef enum
{
    //文件结束
    ENDFILE,
```

```
    //错误
    ERROR,

    //关键字
    IF,             //if
    THEN,           //then
    ELSE,           //else
    END,            //end
    REPEAT,         //repeat
    UNTIL,          //until
    READ,           //read
    WRITE,          //write

    //标识符
    ID,

    //无符号整数
    NUM,

    //特殊符号
    ASSIGN,         //:=
    EQ,             //=
    LT,             //<
    PLUS,           //+
    MINUS,          //-
    TIMES,          // *
    OVER,           ///
    LPAREN,         //(
    RPAREN,         //)
    SEMI,           //;

    //注释
    COMMENT         //{...}

}TokenType;

#endif                  //_DEFINE_H_
```

源代码清单 3-3：scan.txt 文件

```
%{

#include<stdio.h>
#include "define.h"
```

```
int lineno=0;              //行数

//符号计数器
int error_no=0;

int if_no=0;
int then_no=0;
int else_no=0;
int end_no=0;
int repeat_no=0;
int until_no=0;
int read_no=0;
int write_no=0;

int id_no=0;
int num_no=0;

int assign_no=0;
int eq_no=0;
int lt_no=0;
int plus_no=0;
int minus_no=0;
int times_no=0;
int over_no=0;
int lparen_no=0;
int rparen_no=0;
int semi_no=0;

int comment_no=0;

%}

newline                 \n
whitespace              [\t]+

%%

":="                    { return ASSIGN; }
"="                     { return EQ; }
"<"                     { return LT; }
"+"                     { return PLUS; }
"-"                     { return MINUS; }
"*"                     { return TIMES; }
"/"                     { return OVER; }
```

```
"("                    { return LPAREN; }
")"                    { return RPAREN; }
";"                    { return SEMI; }

{newline}              { lineno++; }
{whitespace}           { /*忽略空白*/ }

"{"                    {
                           //匹配注释{...}
                           char c;
                           int comment=1;
                           do
                           {
                               c=input();
                               if(c==EOF)
                               {
                                   comment=0;
                                   break;
                               }
                               else if(c=='\n')
                                   lineno++;
                               else if(c=='}')
                                   break;
                           }while(1);

                           return comment ? COMMENT : ERROR;
                       }

.                      { return ERROR; }

%%

TokenType id2keyword(const char * token);
void stat(TokenType tt, const char * token);
void output();

/*
功能：
    主函数。

参数：
    argc-argv 数组的长度,大小至少为1,argc-1 为命令行参数的数量。
```

argv-字符串指针数组,数组长度为命令行参数个数+1。其中,argv[0]固定指向当前所执行的可执行文件的路径字符串,argv[1]及后面的指针指向各个命令行参数。例如,通过命令行输入"C:\hello.exe-a-b"后,main 函数的 argc 的值为 3,argv[0]指向字符串"C:\hello.exe",argv[1]指向字符串"-a",argv[2]指向字符串"-b"。

返回值:
  成功返回 0,失败返回 1。
*/

```
int main(int argc, char * argv[])
{
    TokenType tt;

    //使用第一个参数输入待处理文件的名称,若没有输入此参数,就报告错误
    if(argc<2)
    {
        printf("Usage: scan.exe filename.\n");
        return 1;
    }

    //打开待处理的文件
    FILE * file=fopen(argv[1], "rt");
    if(NULL==file)
    {
        printf("Can not open file \"%s\".\n", argv[1]);
        return 1;
    }

    //将打开的文件作为 lex 扫描程序的输入
    yyin=file;

    //开始扫描,直到文件结束
    while((tt=yylex())!=ENDFILE)
    {
        //根据符号类型统计其数量
        stat(tt, yytext);
    }

    //输出统计结果
    output();

    //关闭文件
    fclose(file);
```

```c
        return 0;
}

//定义关键字与其类型的映射关系
typedef struct _KeyWord_Entry
{
    const char * word;
    TokenType type;
}KeyWord_Entry;

static const KeyWord_Entry key_table[]=
{
    { "if",     IF      },
    { "then",   THEN    },
    { "else",   ELSE    },
    { "end",    END     },
    { "repeat", REPEAT  },
    { "until",  UNTIL   },
    { "read",   READ    },
    { "write",  WRITE   }
};

/*
功能：
    将标识符转换为对应的关键字类型。

参数：
    id-标识符字符串指针。可能是一个关键字,也可能是用户定义的标识符。

返回值：
    成功返回 0,失败返回 1。
*/
TokenType id2keyword(const char * id)
{
    //
    //TODO：在此添加源代码
    //

    return ID;
}

/*
功能：
```

根据符号类型进行数量统计。

参数：
    tt-符号类型。
    token-符号字符串指针。当符号被识别为标识符时,需要判断其是否为一个关键字。

返回值：
    空。
*/

```c
void stat(TokenType tt, const char * token)
{
    if(ID==tt)
    {
        tt=id2keyword(token);
    }

    switch(tt)
    {
    case IF:         //if
        if_no++;
        break;
    case THEN:       //then
        then_no++;
        break;
    case ELSE:       //else
        else_no++;
        break;
    case END:        //end
        end_no++;
        break;
    case REPEAT:     //repeat
        repeat_no++;
        break;
    case UNTIL:      //until
        until_no++;
        break;
    case READ:       //read
        read_no++;
        break;
    case WRITE:      //write
        write_no++;
        break;
    case ID:         //标识符
```

```
            id_no++;
            break;
        case NUM:          //无符号整数
            num_no++;
            break;
        case ASSIGN:       //:=
            assign_no++;
            break;
        case EQ:           //=
            eq_no++;
            break;
        case LT:           //<
            lt_no++;
            break;
        case PLUS:         //+
            plus_no++;
            break;
        case MINUS:        //-
            minus_no++;
            break;
        case TIMES:        // *
            times_no++;
            break;
        case OVER:         // /
            over_no++;
            break;
        case LPAREN:       //(
            lparen_no++;
            break;
        case RPAREN:       //)
            rparen_no++;
            break;
        case SEMI:         //;
            semi_no++;
            break;
        case COMMENT:      //{...}
            comment_no++;
            break;
        case ERROR:        //错误
            error_no++;
            break;
        }
    }
```

```c
//输出统计结果
void output()
{
    printf("if: %d\n", if_no);
    printf("then: %d\n", then_no);
    printf("else: %d\n", else_no);
    printf("end: %d\n", end_no);
    printf("repeat: %d\n", repeat_no);
    printf("until: %d\n", until_no);
    printf("read: %d\n", read_no);
    printf("write: %d\n", write_no);

    printf("id: %d\n", id_no);
    printf("num: %d\n", num_no);

    printf("assign: %d\n", assign_no);
    printf("eq: %d\n", eq_no);
    printf("lt: %d\n", lt_no);
    printf("plus: %d\n", plus_no);
    printf("minus: %d\n", minus_no);
    printf("times: %d\n", times_no);
    printf("over: %d\n", over_no);
    printf("lparen: %d\n", lparen_no);
    printf("rparen: %d\n", rparen_no);
    printf("semi: %d\n", semi_no);

    printf("comment: %d\n", comment_no);
    printf("error: %d\n", error_no);
    printf("line: %d\n", lineno);
}
```

3.3 生成项目

按照下面的步骤生成项目：

(1) 在"生成"菜单中选择"重新生成项目"（快捷键是 Ctrl＋Alt＋F7）。

(2) 在生成的过程中，CP Lab 会首先使用 Flex 程序根据输入文件 scan.txt 来生成 main.c 文件，然后，将 main.c 文件重新编译、链接为可以运行的可执行文件。

(3) 在生成的 main.c 文件中，尝试找到 scan.txt 文件中第一部分和第三部分 C 源代码插入的位置，并尝试查找 input 函数和 yylex 函数的定义，以及 yyin 和 yytext 等变量的定义。

注意：在下面的实验步骤中，如果需要生成项目，应尽量使用"重新生成项目"功能。如果习惯使用"生成项目"（快捷键是 F7）功能，可能需要连续使用两次此功能才能生成最新的项目。

3.4 运行项目

在没有对项目的源代码进行任何修改的情况下,按照下面的步骤运行项目:

(1) 选择"调试"菜单中的"开始执行(不调试)"(快捷键是 Ctrl+F5)。

(2) 在启动运行时,CP Lab 会自动将 sample.txt 文件的名称作为参数传给可执行文件(即 main 函数中 argv[1]指向的字符串),所以,程序运行完毕后,会在 Windows 控制台窗口中显示对 sample.txt 文件的扫描结果,如图 3-1 所示。

图 3-1 对 sample.txt 文件的扫描结果

由于此时还没有在 scan.txt 中添加能够与标识符和正整数匹配的正则表达式,所以 id 和 num 的值均为 0,而标识符和正整数会被匹配为错误,所以,error 的数量大于 0。

3.5 添加标识符和正整数的统计功能

按照下面的步骤完成此练习:

(1) 在 scan.txt 文件的第一部分添加标识符和正整数的正则表达式。

(2) 在 scan.txt 文件的第二部分添加标识符和正整数的正则表达式匹配时的 C 源代码。注意,标识符的正则表达式也用来匹配所有的关键字,不要直接使用字符串来逐个匹配关键字。

(3) 重新生成项目。如果生成失败,根据"输出"窗口中的提示信息修改源代码中的语法错误。

(4) 按 Ctrl+F5 键启动执行项目。

执行的结果如图 3-2 所示,已经可以正确统计出标识符和正整数的数量。注意,由于此时 id2keyword 函数还无法将标识符转换为关键字,所以,所有的关键字都匹配成了标识符,关键字的数量都为 0。

图 3-2 正确统计出标识符和正整数的数量

3.6 添加关键字的统计功能

按照下面的步骤完成此练习：

（1）为 id2keyword 函数编写源代码。要求使用此函数前面定义的 key_table 表格中的数据，通过线性搜索的方式，根据标识符的字符串确定其对应的关键字类型。

（2）重新生成项目。如果生成失败，根据"输出"窗口中的提示信息修改源代码中的语法错误。

（3）按 Ctrl+F5 键启动执行项目。

执行的结果如图 3-3 所示，已经可以正确统计出各个关键字的数量。

图 3-3 正确统计出各个关键字的数量

注意：

（1）本实验的模板不提供演示功能，所以，如果执行的结果不正确，可以通过添加断点和单步调试的方法来查找错误的原因。

（2）断点应该添加在 scan.txt 文件中需要中断的 C 源代码行，不要添加在 main.c 文件中，否则无法命中断点。

3.7　添加 C 语言风格的注释

按照下面的步骤完成此练习：

（1）将 sample.txt 文件中的多行注释包括在"/*"和"*/"之间，将语句末尾的注释放在"//"的后面。

（2）修改 scan.txt 文件，使之能够正确匹配和统计这两种 C 语言风格的注释。

（3）重新生成项目。如果生成失败，根据"输出"窗口中的提示信息修改源代码中的语法错误。

（4）按 Ctrl＋F5 键启动执行项目，确保统计的注释数量是正确的。

四、思考与练习

1. 修改 key_table 表格中数据的顺序，使关键字按照字母顺序排列，然后修改 id2keyword 函数中的代码，使用二分法从 key_table 表格中查找关键字。

2. 使用 gperf 工具为 TINY 语言的关键字生成杂凑表（哈希表），然后修改 id2keyword 函数中的代码从杂凑表中查找关键字。（提示：CP Lab 提供了 gperf 工具，可以选择 CP Lab"工具"菜单中的"CP Lab 命令提示"，在弹出的 Windows 控制台窗口中输入命令"gperf C:\a.txt --output-file＝C:\a.c"，即可为 a.txt 文件中列出的关键字生成杂凑表到 a.c 源代码文件中。选择 CP Lab"帮助"菜单中"其他帮助文档"中的"gperf 手册"可以获得更多帮助。）

3. 选择 CP Lab"帮助"菜单中"其他帮助文档"中的"Flex 手册"，学习 Flex 工具的更多用法。如果需要修改 Flex 程序在处理输入文件时的选项，可以在"项目管理器"窗口中右击文件 scan.txt，在弹出的快捷菜单中选择"属性"，然后选择"属性页"左侧"自定义生成步骤"的"常规"，就可以编辑右侧的"命令行"选项了。

EXPERIMENT 4

实验 4　消除左递归(无替换)

实验难度：★★★☆☆
建议学时：2学时

一、实验目的

- 了解在上下文无关文法中的左递归的概念。
- 掌握直接左递归的消除算法。

二、预备知识

- 在这个实验中用到了单链表插入和删除操作。如果读者对这一部分知识有遗忘，可以复习一下数据结构中的相关内容。
- 理解指针的指针的概念和用法。在这个实验中，指针的指针被用来确定单链表插入的位置，以及插入的具体操作。
- 理解左递归和右递归的含义以及左递归的各种形式，如简单直接左递归、普遍的直接左递归、一般的左递归。

三、实验内容

3.1　准备实验

按照下面的步骤准备实验：

(1) 启动 CP Lab。
(2) 在"文件"菜单中选择"新建"|"项目"，打开"新建项目"对话框。
(3) 使用模板"004 消除左递归(无替换)"新建一个项目。

3.2　阅读实验源代码

1. RemoveLeftRecursion.h 文件(参见源代码清单 4-1)

此文件主要定义了与文法相关的数据结构，这些数据结构定义了文法的单链表存储形式。其中 Rule 结构体用于定义文法的名称和文法链表；RuleSymbol 结构体用于定义文法产生式中的终结符和非终结符。具体内容可参见表 4-1 和表 4-2。

表 4-1

Rule 的域	说　　明
RuleName	文法的名称
pFirstSymbol	指向文法的第一个 Select 的第一个 Symbol
pNextRule	指向下一条文法

表 4-2

RuleSymbol 的域	说　　明
pNextSymbol	指向下一个 Symbol
pOther	指向下一个 Select
isToken	是否为终结符。1 表示终结符,0 表示非终结符
TokenName	终结符的名称。isToken 为 1 时这个域有效
pRule	指向 Symbol 对应的 Rule。isToken 为 0 时这个域有效

下面是一个简单文法,并使用图 4-1 和图 4-2 说明了该文法的存储结构:

A->Aa|aB
B->bB

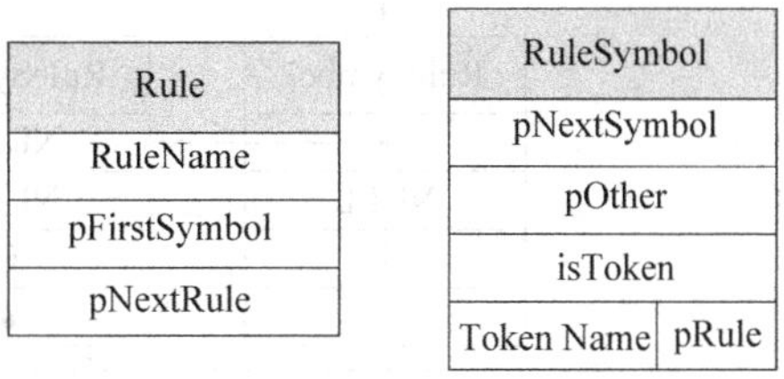

图 4-1　Rule 和 RuleSymbol 结构体图例

2. main.c 文件(参见源代码清单 4-2)

此文件定义了 main 函数。在 main 函数中首先调用 InitRules 函数初始化了文法,然后调用 PrintRule 函数打印消除左递归之前的文法,接着调用 RemoveLeftRecursion 函数对文法消除左递归,最后再次调用 PrintRule 函数打印消除左递归之后的文法。

在 main 函数的后面定义了一系列函数,有关这些函数的具体内容参见表 4-3。关于这些函数的参数和返回值,可以参见其注释。

表 4-3

函　数　名	功　能　说　明
AddSymbolToSelect	将一个 Symbol 添加到 Select 的末尾。此函数的函数体还不完整,留给读者完成
AddSelectToRule	将 Select 加入到文法末尾,如果 Select 为 NULL 则将 ε 终结符加入到文法末尾。在本程序中 ε 可以用 $ 来代替。此函数的函数体还不完整,留给读者完成

实验 4　消除左递归(无替换)

续表

函 数 名	功 能 说 明
RemoveLeftRecursion	对文法消除左递归。在本函数中使用指针的指针 pSelectPtr 来确定符号在单链表中插入和删除的位置,在进入下一次循环之前应为 pSelectPtr 设置正确的值。此函数的函数体还不完整,留给读者完成
InitRules	使用给定的数据初始化文法链表
CreateRule	创建一个新的 Rule
CreateSymbol	创建一个新的 Symbol
FindRule	根据 RuleName 在文法链表中查找名字相同的文法
PrintRule	输出文法。此函数的函数体还不完整,留给读者完成

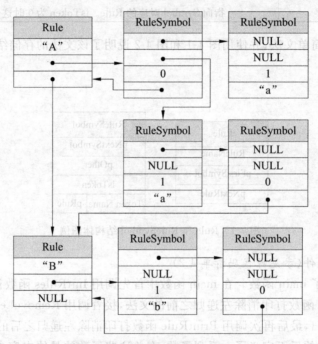

图 4-2 简单文法的存储结构

源代码清单 4-1:RemoveLeftRecursion.h 文件

```
#ifndef _REMOVELEFTRECURSIONNOREPLACE_H_
#define _REMOVELEFTRECURSIONNOREPLACE_H_

//
//在此处包含 C 标准库头文件
//

#include<stdio.h>
```

```
//
//在此处包含其他头文件
//
//
//在此处定义数据结构
//

#define MAX_STR_LENGTH 64

struct _Rule;
typedef struct _RuleSymbol{
    struct _RuleSymbol * pNextSymbol;   //指向下一个Symbol
    struct _RuleSymbol * pOther;        //指向下一个Select
    int isToken;                        //是否为终结符。1表示终结符,0表示非终结符
    char TokenName[MAX_STR_LENGTH];
                                        //终结符的名称。isToken为1时这个域有效
    struct _Rule * pRule;
                                        //指向Symbol对应的Rule。isToken为0时这个域有效
}RuleSymbol;

typedef struct _Rule{
    char RuleName[MAX_STR_LENGTH];      //文法的名称
    struct _RuleSymbol * pFirstSymbol;  //指向文法的第一个Select的第一个Symbol
    struct _Rule * pNextRule;           //指向下一条文法
}Rule;

//
//在此处声明函数
//

Rule * InitRules();
Rule * CreateRule(const char * pRuleName);
RuleSymbol * CreateSymbol();
Rule * FindRule(Rule * pHead, const char * RuleName);

void AddSymbolToSelect(RuleSymbol * pSelect, RuleSymbol * pNewSymbol);
void AddSelectToRule(Rule * pRule, RuleSymbol * pNewSelect);
void RemoveLeftRecursion(Rule * pHead);

void PrintRule(Rule * pHead);

//
//在此处声明全局变量
```

```
//
extern const char * VoidSymbol;
extern const char * Postfix;


#endif /* _REMOVELEFTRECURSIONNOREPLACE_H_ */
```

源代码清单 4-2: `main.c` 文件

```c
#include "RemoveLeftRecursion.h"

const char * VoidSymbol="$ "; //"ε"
const char * Postfix="'";

int main(int argc, char * argv[])
{
    //
    //调用 InitRules 函数初始化文法
    //
    Rule * pHead=InitRules();

    //
    //输出消除左递归之前的文法
    //
    printf("Before Remove Left Recursion:\n");
    PrintRule(pHead);

    //
    //调用 RemoveLeftRecursion 函数对文法消除左递归
    //
    RemoveLeftRecursion(pHead);

    //
    //输出消除左递归之后的文法
    //
    printf("\nAfter Remove Left Recursion:\n");
    PrintRule(pHead);

    return 0;
}

/*
功能:
```

```
    将一个 Symbol 添加到 Select 的末尾。

参数:
    pSelect--Select 指针。
    pNewSymbol--Symbol 指针。
*/
void AddSymbolToSelect(RuleSymbol * pSelect, RuleSymbol * pNewSymbol)
{

    //
    //TODO: 在此处添加代码
    //

}

/*
功能:
    将一个 Select 加入到文法末尾。当 Select 为 NULL 时就将一个 ε 终结符加入到文法末尾。

参数:
    pRule--文法指针。
    pNewSelect--Select 指针。
*/
void AddSelectToRule(Rule * pRule, RuleSymbol * pNewSelect)
{

    //
    //TODO: 在此处添加代码
    //

}

/*
功能:
    消除左递归。

参数:
    pHead--文法链表的头指针。
*/
void RemoveLeftRecursion(Rule * pHead)
{
    RuleSymbol * pSelect;           //Select 游标
```

实验4　消除左递归(无替换)

```
    Rule * pNewRule;                    //Rule 指针

    //
    //TODO: 在此处添加代码
    //

    return;
}

/*
功能:
    使用给定的数据初始化文法链表。

返回值:
    Rule 指针。
*/
typedef struct _SYMBOL{
    int isToken;
    char Name[MAX_STR_LENGTH];
}SYMBOL;

typedef struct _RULE_ENTRY{
    char RuleName[MAX_STR_LENGTH];
    SYMBOL Selects[64][64];
}RULE_ENTRY;

static const RULE_ENTRY rule_table[]=
{
    /* A->Aa | bA | c | Ad */
    { "A", {
            { { 0, "A" }, { 1, "a" } },
            { { 1, "b" }, { 0, "A" } },
            { { 1, "c" } },
            { { 0, "A" }, { 1, "d" } }
        }
    }
};

Rule * InitRules()
{
    Rule * pHead, * pRule;
    RuleSymbol * * pSymbolPtr1, * * pSymbolPtr2;
    int nRuleCount=sizeof(rule_table) / sizeof(rule_table[0]);
```

```c
    int i, j, k;

    Rule * * pRulePtr=&pHead;
    for(i=0; i<nRuleCount; i++)
    {
        * pRulePtr=CreateRule(rule_table[i].RuleName);
        pRulePtr=&(* pRulePtr)->pNextRule;
    }

    pRule=pHead;
    for(i=0; i<nRuleCount; i++)
    {
        pSymbolPtr1=&pRule->pFirstSymbol;
        for(j=0; rule_table[i].Selects[j][0].Name[0]!='\0'; j++)
        {
            pSymbolPtr2=pSymbolPtr1;
            for(k=0; rule_table[i].Selects[j][k].Name[0]!='\0'; k++)
            {
                const SYMBOL * pSymbol=&rule_table[i].Selects[j][k];

                * pSymbolPtr2=CreateSymbol();
                (* pSymbolPtr2)->isToken=pSymbol->isToken;
                if(1==pSymbol->isToken)
                {
                    strcpy((* pSymbolPtr2)->TokenName, pSymbol->Name);
                }
                else
                {
                    (* pSymbolPtr2)->pRule=FindRule(pHead, pSymbol->Name);
                    if(NULL==(* pSymbolPtr2)->pRule)
                    {
                        printf("Init rules error, miss rule \"%s\"\n", pSymbol
                            ->Name); exit(1);
                    }
                }

                pSymbolPtr2=&(* pSymbolPtr2)->pNextSymbol;
            }

            pSymbolPtr1=&(* pSymbolPtr1)->pOther;
        }

        pRule=pRule->pNextRule;
```

```
        }

        return pHead;
}

/*
功能：
    创建一个新的 Rule。

参数：
    pRuleName--文法的名字。

返回值：
    Rule 指针。
*/
Rule * CreateRule(const char * pRuleName)
{
    Rule * pRule=(Rule * )malloc(sizeof(Rule));

    strcpy(pRule->RuleName, pRuleName);
    pRule->pFirstSymbol=NULL;
    pRule->pNextRule=NULL;

    return pRule;
}

/*
功能：
    创建一个新的 Symbol。

返回值：
    RuleSymbol 指针。
*/
RuleSymbol * CreateSymbol()
{
    RuleSymbol * pSymbol=(RuleSymbol * )malloc(sizeof(RuleSymbol));

    pSymbol->pNextSymbol=NULL;
    pSymbol->pOther=NULL;
    pSymbol->isToken=-1;
    pSymbol->TokenName[0]='\0';
    pSymbol->pRule=NULL;

    return pSymbol;
}
```

```
/*
功能：
    根据 RuleName 在文法链表中查找名字相同的文法。

参数：
    pHead--文法链表的头指针。
    RuleName--文法的名字。

返回值：
    如果存在名字相同的文法,返回 Rule 指针,否则返回 NULL。
*/
Rule * FindRule(Rule * pHead, const char * RuleName)
{
    Rule * pRule;
    for(pRule=pHead; pRule!=NULL; pRule=pRule->pNextRule)
    {
        if(0==strcmp(pRule->RuleName, RuleName))
        {
            break;
        }
    }

    return pRule;
}

/*
功能：
    输出文法。

参数：
    pHead--文法链表的头指针。
*/
void PrintRule(Rule * pHead)
{

    //
    //TODO: 在此处添加代码
    //

}
```

3.3 为函数 InitRules 添加注释

在 InitRules 函数之前定义了两个结构体,这两个结构体用来定义初始化文法数据的存储形式。具体内容可参见表 4-4 和表 4-5。

表 4-4

SYMBOL 的域	说 明
isToken	是否为终结符。1 表示终结符,0 表示非终结符
Name	终结符和非终结符的名称

表 4-5

RULE_ENTRY 的域	说 明
RuleName	文法的名称
Selects	SYMBOL 结构体的二维数组,其中每一行表示一个 Select,一行中的每个元素分别表示一个终结符或非终结符

为 InitRules 函数添加注释(注意,在本程序中使用了指针的指针,体会其在单链表的插入和删除操作中的作用)。

3.4 为 PrintRule 函数编写源代码

为 PrintRule 函数编写源代码,同时理解在本程序中文法的链式存储结构。编写完源代码后,选择"调试"菜单中的"调试/开始执行(不调试)",会在控制台窗口输出文法的产生式,如图 4-3 所示。由于还没有为 RemoveLeftRecursion 函数和其他未完成的函数编写源代码,所以前后两次输出的文法是一样的。

图 4-3 打印文法

3.5 在演示模式下调试项目

按照下面的步骤调试项目:

(1) 按 F7 键生成项目。
(2) 在演示模式下,按 F5 键启动调试项目。程序会在观察点函数的开始位置中断。
(3) 重复按 F5 键,直到调试过程结束。

在调试的过程中,每执行"演示流程"窗口中的一行后,仔细观察"转储信息"窗口内容所发生的变化,理解对文法消除左递归的过程。"转储信息"窗口显示的数据信息包括:

- 文法。由于在本实验中没有进行替换操作,所以只能处理一条文法的情况。
- 新文法。对原文法消除左递归后,新生成的包含右递归的文法。
- 执行到函数末尾时,会将新文法加入到文法链表中。
- 在消除左递归的过程中会使用加粗的中括号表示出游标指向的 Select。

3.6 编写源代码并通过验证

按照下面的步骤继续实验：

（1）为 RemoveLeftRecursion 函数和其他未完成的函数编写源代码。注意尽量使用已定义的局部变量。

（2）按 F7 键生成项目。如果生成失败，根据"输出"窗口中的提示信息修改源代码中的语法错误。

（3）按 Alt＋F5 键启动验证。如果验证失败，可以使用"输出"窗口中的"比较"功能，或者在"非演示模式"下按 F5 键启动调试后重复按 F10 键单步调试读者编写的源代码，从而定位错误的位置，然后回到步骤(1)。

四、思考与练习

1. 请读者仔细检查自己编写的源代码，查看在消除左递归的过程中是否调用了 free 函数释放了从文法链表中移除的 Symbol，否则会造成内存泄露。

2. 编写一个 FreeRule 函数，当在 main 函数的最后调用此函数时，可以将整个文法的内存释放掉，从而避免内存泄露。

3. 本实验目前要求读者编写的代码只能够处理一条包含直接左递归的文法，请读者思考一下，如果有多条文法都包含左递归，甚至还包含了间接左递归，应该如何改进现有的源代码。

EXPERIMENT 5

实验 5　消除左递归（有替换）

实验难度：★★★★☆
建议学时：4 学时

一、实验目的

- 了解在上下文无关文法中的左递归的概念。
- 掌握直接左递归、一般左递归的消除算法。

二、预备知识

- 在这个实验中用到了单链表插入和删除操作。如果读者对这一部分知识有遗忘，可以复习一下数据结构中的相关内容。
- 理解指针的指针的概念和用法。在这个实验中，指针的指针被用来确定单链表插入和删除的位置，以及用来完成插入和删除的具体操作。
- 理解左递归和右递归的含义以及左递归的各种形式，如简单直接左递归、普遍的直接左递归、一般的左递归。

三、实验内容

3.1　准备实验

按照下面的步骤准备实验：
（1）启动 CP Lab。
（2）在"文件"菜单中选择"新建"|"项目"，打开"新建项目"对话框。
（3）使用模板"005 消除左递归（有替换）"新建一个项目。

3.2　阅读实验源代码

1. RemoveLeftRecursion.h 文件（参见源代码清单 5-1）

此文件主要定义了与文法相关的数据结构，这些数据结构定义了文法的单链表存储形式。其中，Rule 结构体用于定义文法的名称和文法链表；RuleSymbol 结构体用于定义文法产生式中的终结符和非终结符。具体内容可参见表 5-1 和表 5-2。

表 5-1

Rule 的域	说明
RuleName	文法的名称
pFirstSymbol	指向文法的第一个 Select 的第一个 Symbol
pNextRule	指向下一条文法

表 5-2

RuleSymbol 的域	说明
pNextSymbol	指向下一个 Symbol
pOther	指向下一个 Select
isToken	是否为终结符。1 表示终结符，0 表示非终结符
TokenName	终结符的名称。isToken 为 1 时这个域有效
pRule	指向 Symbol 对应的 Rule。isToken 为 0 时这个域有效

下面是一个简单文法，并使用图 5-1 和 5-2 说明了该文法的存储结构：

```
A->Aa|aB
B->bB
```

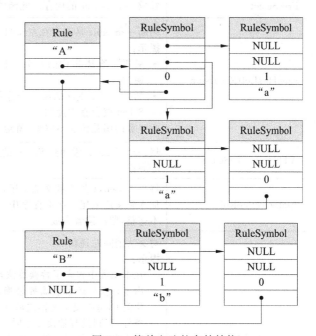

图 5-1　Rule 和 RuleSymbol 结构体图例　　　图 5-2　简单文法的存储结构

2. main.c 文件（参见源代码清单 5-2）

此文件定义了 main 函数。在 main 函数中首先调用 InitRules 函数初始化了文法，然后调用 PrintRule 函数打印消除左递归之前的文法，接着调用 RemoveLeftRecursion 函数对文法消除左递归，最后再次调用 PrintRule 函数打印消除左递归之后的文法。

在 main 函数的后面定义了一系列函数，有关这些函数的具体内容参见表 5-3。关于

实验 5　消除左递归（有替换）

这些函数的参数和返回值,可以参见其注释。

表 5-3

函 数 名	功 能 说 明
SymbolNeedReplace	判断当前 Rule 中的一个 Symbol 是否需要被替换。如果 Symbol 是一个非终结符,且 Symbol 对应的 Rule 在当前 Rule 之前,就需要被替换。此函数的函数体还不完整,留给读者完成
CopySymbol	复制一个 Symbol。此函数的函数体还不完整,留给读者完成
CopySelect	复制一个 Select。在此函数中可以调用 CopySymbol 函数,将 Select 中的 Symbol 逐个进行复制。此函数的函数体还不完整,留给读者完成
ReplaceSelect	替换一个 Select 的第一个 Symbol 提示: • 在调用此函数之前,已经调用了 SymbolNeedReplace 函数,确保 Select 的第一个 Symbol 需要被替换 • Select 的第一个 Symbol 是一个非终结符,该 Symbol 对应的 Rule 就是用于替换该 Symbol 模板 • 当需要复制 Select 时,可以调用 CopySelect 函数。 此函数的函数体还不完整,留给读者完成
FreeSelect	释放一个 Select 的内存。此函数的函数体还不完整,留给读者完成
RuleHasLeftRecursion	判断一条 Rule 是否存在左递归 提示: • 只要一条 Rule 中的一个 Select 存在左递归,那么这条 Rule 就存在左递归 • 如果一个 Select 的第一个符号(非终结符)对应的就是此条文法,这个 Select 就存在左递归 此函数的函数体还不完整,留给读者完成
AddSymbolToSelect	将一个 Symbol 添加到 Select 的末尾。此函数的函数体还不完整,留给读者完成
AddSelectToRule	将一个 Select 加入到文法末尾,当 Select 为 NULL 时就将一个 ε 终结符加入到文法末尾。在本程序中 ε 可以用 $ 来代替。此函数的函数体还不完整,留给读者完成
RemoveLeftRecursion	对文法消除左递归 提示: • 在本函数中包含了替换算法,从而可以处理间接左递归的情况。而且由于替换后还可能需要替换,所以设置了一个标识 isChange,并初始化为 0,每当发生替换之后就将该标识赋值为 1,并将此标识作为循环条件,直到没有替换发生时,才能结束替换 • 在本函数中使用指针的指针 pSelectPtr 来确定符号在单链表中插入和删除的位置,在进入下一次循环之前应为 pSelectPtr 设置正确的值 • 在每处理完一条文法后,会将文法链表的游标指向新文法,这样在继续执行 for 循环的移动游标操作后,就会跳过这条新文法,从而继续对后面的文法进行消除左递归操作(新文法不包含左递归,所以并不需要对这条文法进行处理) 此函数的函数体还不完整,留给读者完成

续表

函 数 名	功 能 说 明
InitRules	使用给定的数据初始化文法链表
CreateRule	创建一个新的 Rule
CreateSymbo	创建一个新的 Symbol
FindRule	根据 RuleName 在文法链表中查找名字相同的文法
PrintRule	输出文法。此函数的函数体还不完整,留给读者完成

源代码清单 5-1: RemoveLeftRecursion.h 文件

```c
#ifndef _REMOVELEFTRECURSION_H_
#define _REMOVELEFTRECURSION_H_

//
//在此处包含 C 标准库头文件
//

#include<stdio.h>

//
//在此处包含其他头文件
//

//
//在此处定义数据结构
//

#define MAX_STR_LENGTH 64

struct _Rule;
typedef struct _RuleSymbol{
    struct _RuleSymbol * pNextSymbol; //指向下一个 Symbol
    struct _RuleSymbol * pOther;      //指向下一个 Select
    int isToken;                      //是否为终结符。1表示终结符,0表示非终结符
    char TokenName[MAX_STR_LENGTH];   //终结符的名称。isToken 为 1 时这个域有效
    struct _Rule * pRule;             //指向 Symbol 对应的 Rule。isToken 为 0 时这个域有效
}RuleSymbol;

typedef struct _Rule{
    char RuleName[MAX_STR_LENGTH];    //文法的名称
    struct _RuleSymbol * pFirstSymbol; //指向文法的第一个 Select 的第一个 Symbol
```

实验 5 消除左递归(有替换)

```c
    struct _Rule * pNextRule;           //指向下一条文法
}Rule;

//
//在此处声明函数
//
Rule * InitRules();
Rule * CreateRule(const char * pRuleName);
RuleSymbol * CreateSymbol();
Rule * FindRule(Rule * pHead, const char * RuleName);

int RuleHasLeftRecursion(Rule * pRule);
void AddSymbolToSelect(RuleSymbol * pSelect, RuleSymbol * pNewSymbol);
void AddSelectToRule(Rule * pRule, RuleSymbol * pNewSelect);
RuleSymbol * CopySymbol(const RuleSymbol * pSymbolTemplate);
RuleSymbol * CopySelect(const RuleSymbol * pSelectTemplate);
void FreeSelect(RuleSymbol * pSelect);
RuleSymbol * ReplaceSelect(const RuleSymbol * pSelectTemplate);
void RemoveLeftRecursion(Rule * pHead);

void PrintRule(Rule * pHead);

//
//在此处声明全局变量
//

extern const char * VoidSymbol;
extern const char * Postfix;

#endif  /* _REMOVELEFTRECURSION_H_ */
```

源代码清单 5-2: **main.c** 文件

```c
#include "RemoveLeftRecursion.h"

const char * VoidSymbol="$ ";           //"ε"
const char * Postfix="'";

int main(int argc, char * argv[])
{
    //
    //调用 InitRules 函数初始化文法
    //
    Rule * pHead=InitRules();
```

```
    //
    //输出消除左递归之前的文法
    //
    printf("Before Remove Left Recursion:\n");
    PrintRule(pHead);

    //
    //调用 RemoveLeftRecursion 函数消除文法中的左递归
    //
    RemoveLeftRecursion(pHead);

    //
    //输出消除左递归之后的文法
    //
    printf("\nAfter Remove Left Recursion:\n");
    PrintRule(pHead);

    return 0;
}

/*
功能:
    判断当前 Rule 中的一个 Symbol 是否需要被替换。
    如果 Symbol 是一个非终结符,且 Symbol 对应的
    Rule 在当前 Rule 之前,就需要被替换。

参数:
    pCurRule--当前 Rule 的指针。
    pSymbol--Symbol 指针。

返回值:
    需要替换返回 1。
    不需要替换返回 0。
*/
int SymbolNeedReplace(const Rule * pCurRule, const RuleSymbol * pSymbol)
{

    //
    //TODO:在此添加代码
    //


}
```

实验 5 消除左递归(有替换)

```
/*
功能：
    复制一个Symbol。

参数：
    pSymbolTemplate--需要被复制的Symbol指针。

返回值：
    复制获得的新Symbol的指针。
*/
RuleSymbol * CopySymbol(const RuleSymbol * pSymbolTemplate)
{

    //
    //TODO：在此处添加代码
    //

}


/*
功能：
    复制一个Select。

参数：
    pSelectTemplate--需要被复制的Select指针。

返回值：
    复制获得的新Select的指针。
*/
RuleSymbol * CopySelect(const RuleSymbol * pSelectTemplate)
{

    //
    //TODO：在此处添加代码
    //

}


/*
功能：
    替换一个Select的第一个Symbol。

参数：
    pSelectTemplate--需要被替换的Select指针。
```

返回值:
    替换后获得的新 Select 的指针。
    注意,替换后可能会有一个新的 Select,
    也可能会有多个 Select 链接在一起。
*/
RuleSymbol * ReplaceSelect(const RuleSymbol * pSelectTemplate)
{

    //
    //TODO:在此处添加代码
    //

}

/*
功能:
    释放一个 Select 的内存。

参数:
    pSelect--需要释放的 Select 的指针。
*/
void FreeSelect(RuleSymbol * pSelect)
{

    //
    //TODO:在此处添加代码
    //

}

/*
功能:
    判断一条 Rule 是否存在左递归。

参数:
    prRule--Rule 指针。

返回值:
    存在返回 1。
    不存在返回 0。
*/
int RuleHasLeftRecursion(Rule * pRule)
{

```
    //
    //TODO:在此处添加代码
    //

}

/*
功能:
    将一个 Symbol 添加到 Select 的末尾。

参数:
    pSelect--Select 指针。
    pNewSymbol--Symbol 指针。
*/
void AddSymbolToSelect(RuleSymbol * pSelect, RuleSymbol * pNewSymbol)
{

    //
    //TODO:在此处添加代码
    //

}

/*
功能:
    将一个 Select 加入到文法末尾,当 Select 为 NULL 时就将一个 ε 终结符加入到文法末尾。

参数:
    pRule--文法指针。
    pNewSelect--Select 指针。
*/
void AddSelectToRule(Rule * pRule, RuleSymbol * pNewSelect)
{

    //
    //TODO:在此处添加代码
    //

}

/*
功能:
    消除左递归。
```

参数:
    pHead--文法链表的头指针。
*/
```c
void RemoveLeftRecursion(Rule * pHead)
{
    Rule * pRule;                        //Rule 游标
    RuleSymbol * pSelect;                //Select 游标
    Rule * pNewRule;                     //Rule 指针
    int isChange;                        //Rule 是否被替换的标记
    RuleSymbol * * pSelectPrePtr;        //Symbol 指针的指针

    //
    //TODO: 在此处添加代码
    //

}
```

/*
功能:
    使用给定的数据初始化文法链表。

返回值:
    Rule 指针。
*/
```c
typedef struct _SYMBOL{
    int isToken;
    char Name[MAX_STR_LENGTH];
}SYMBOL;

typedef struct _RULE_ENTRY{
    char RuleName[MAX_STR_LENGTH];
    SYMBOL Selects[64][64];
}RULE_ENTRY;

static const RULE_ENTRY rule_table[]=
{
    /* A->Ba | Aa | c */
    { "A", {
            { { 0, "B" }, { 1, "a"} },
            { { 0, "A" }, { 1, "a"} },
            { { 1, "c" } }
        }
    },
```

```c
        /* B->Bb | Ab | d */
        { "B", {
                { { 0, "B" }, { 1, "b"} },
                { { 0, "A" }, { 1, "b"} },
                { { 1, "d" } }
            }
        }
};

Rule * InitRules()
{
    Rule * pHead, * pRule;
    RuleSymbol * * pSymbolPtr1, * * pSymbolPtr2;
    int nRuleCount=sizeof(rule_table) / sizeof(rule_table[0]);
    int i, j, k;

    Rule * * pRulePtr=&pHead;
    for(i=0; i<nRuleCount; i++)
    {
        *pRulePtr=CreateRule(rule_table[i].RuleName);
        pRulePtr=&(*pRulePtr)->pNextRule;
    }

    pRule=pHead;
    for(i=0; i<nRuleCount; i++)
    {
        pSymbolPtr1=&pRule->pFirstSymbol;
        for(j=0; rule_table[i].Selects[j][0].Name[0]!='\0'; j++)
        {
            pSymbolPtr2=pSymbolPtr1;
            for(k=0; rule_table[i].Selects[j][k].Name[0]!='\0'; k++)
            {
                const SYMBOL * pSymbol=&rule_table[i].Selects[j][k];

                *pSymbolPtr2=CreateSymbol();
                (*pSymbolPtr2)->isToken=pSymbol->isToken;
                if(1==pSymbol->isToken)
                {
                    strcpy((*pSymbolPtr2)->TokenName, pSymbol->Name);
                }
                else
                {
                    (*pSymbolPtr2)->pRule=FindRule(pHead, pSymbol->Name);
```

```
                    if(NULL==(*pSymbolPtr2)->pRule)
                    {
                        printf("Init rules error, miss rule \"%s\"\n", pSymbol->
                        Name);
                        exit(1);
                    }
                }

                pSymbolPtr2=&(*pSymbolPtr2)->pNextSymbol;
            }

            pSymbolPtr1=&(*pSymbolPtr1)->pOther;
        }

        pRule=pRule->pNextRule;
    }

    return pHead;
}

/*
功能:
    创建一个新的 Rule。

参数:
    pRuleName--文法的名字。

返回值:
    Rule 指针。
*/
Rule * CreateRule(const char * pRuleName)
{
    Rule * pRule= (Rule *)malloc(sizeof(Rule));

    strcpy(pRule->RuleName, pRuleName);
    pRule->pFirstSymbol=NULL;
    pRule->pNextRule=NULL;

    return pRule;
}

/*
功能:
```

```
        创建一个新的 Symbol。

返回值:
    RuleSymbol 指针。
*/
RuleSymbol * CreateSymbol()
{
    RuleSymbol * pSymbol=(RuleSymbol *)malloc(sizeof(RuleSymbol));

    pSymbol->pNextSymbol=NULL;
    pSymbol->pOther=NULL;
    pSymbol->isToken=-1;
    pSymbol->TokenName[0]='\0';
    pSymbol->pRule=NULL;

    return pSymbol;
}

/*
功能:
    根据 RuleName 在文法链表中查找名字相同的文法。

参数:
    pHead--文法的头指针。
    RuleName--文法的名字。

返回值:
    Rule 指针。
*/
Rule * FindRule(Rule * pHead, const char * RuleName)
{
    Rule * pRule;
    for(pRule=pHead; pRule!=NULL; pRule=pRule->pNextRule)
    {
        if(0==strcmp(pRule->RuleName, RuleName))
        {
            break;
        }
    }

    return pRule;
}
/*
功能:
    输出文法。
```

```
参数：
    pHead--文法的头指针。
*/
void PrintRule(Rule * pHead)
{
    
    //
    //TODO:在此处添加代码
    //
    
}
```

3.3 为函数 InitRules 添加注释

在 InitRules 函数之前定义了两个结构体，这两个结构体用来定义初始化文法数据的存储形式。具体内容可参见表 5-4 和表 5-5。

表 5-4

SYMBOL 的域	说　　明
isToken	是否为终结符。1 表示终结符，0 表示非终结符
Name	终结符和非终结符的名称

表 5-5

RULE_ENTRY 的域	说　　明
RuleName	文法的名称
Selects	SYMBOL 结构体的二维数组，其中每一行表示一个 Select，一行中的每个元素分别表示一个终结符或非终结符

为 InitRules 函数添加注释（注意，在本程序中使用了指针的指针，体会其在单链表的插入和删除操作中的作用）。

3.4 为 PrintRule 函数编写源代码

为 PrintRule 函数编写源代码，同时理解在本程序中文法的链式存储结构，编写完源代码后，选择"调试"菜单中的"调试/开始执行（不调试）"，会在控制台窗口输出文法的产生式，如图 5-3 所示。由于还没有为 RemoveLeftRecursion 函数和其他未完成的函数编写源代码，所以前后两次输出的文法是一样的。

图 5-3 打印文法

3.5 在演示模式下调试项目

按照下面的步骤调试项目：

实验 5 消除左递归（有替换）

(1) 按 F7 键生成项目。
(2) 在演示模式下，按 F5 键启动调试项目。程序会在观察点函数的开始位置中断。
(3) 重复按 F5 键，直到调试过程结束。

在调试的过程中，每执行"演示流程"窗口中的一行后，仔细观察"转储信息"窗口内容所发生的变化，理解对文法消除左递归的过程。"转储信息"窗口显示的数据信息包括：

- 文法。在消除左递归的过程中会使用游标来指向正在操作的文法。
- 新文法。对原文法消除左递归后，新生成的包含右递归的文法。
- 执行到最外层的循环的末尾时，会将新文法加入到文法链表中。
- 在消除左递归的过程中会使用加粗的中括号表示出游标指向的 Select。

3.6 编写源代码并通过验证

按照下面的步骤继续实验：

(1) 为 RemoveLeftRecursion 函数和其他未完成的函数编写源代码。注意尽量使用已定义的局部变量。

(2) 按 F7 键生成项目。如果生成失败，根据"输出"窗口中的提示信息修改源代码中的语法错误。

(3) 按 Alt+F5 键启动验证。如果验证失败，可以使用"输出"窗口中的"比较"功能，或者在"非演示模式"下按 F5 键启动调试后重复按 F10 键单步调试读者编写的源代码，从而定位错误的位置，然后回到步骤(1)。

四、思考与练习

1. 请读者仔细检查自己编写的源代码，查看在消除左递归的过程中是否调用了 free 函数释放了从文法链表中移除的 Symbol，是否调用了 FreeSelect 函数释放了从文法链表中移除的 Select，否则会造成内存泄露。

2. 编写一个 FreeRule 函数，当在 main 函数的最后调用此函数时，可以将整个文法的内存释放掉，从而避免内存泄露。

3. 使用自己编写的代码对下面两个例子进行验证，确保程序可以为所有形式的文法消除左递归，并验证通过（文法中的终结符用粗体表示）。

例 1：

```
A->Ba|Aa|c
B->Bb|Ab|D
D->Ad
```

例 2：

```
exp->exp addop term|term
addop->+|-
term->term mulop factor|factor
mulop->*
factor->(exp)|number
```

EXPERIMENT 6

实验 6　提取左因子

实验难度：★★★☆☆
建议学时：2 学时

一、实验目的

- 了解在上下文无关文法中的左因子的概念。
- 掌握提取左因子的算法。

二、预备知识

- 在这个实验中用到了单链表插入和删除操作。如果读者对这一部分知识有遗忘，可以复习一下数据结构中的相关内容。
- 理解指针的指针的概念和用法。在这个实验中，指针的指针被用来确定单链表插入和删除的位置，以及用来完成插入和删除的具体操作。

三、实验内容

3.1　准备实验

按照下面的步骤准备实验：
(1) 启动 CP Lab。
(2) 在"文件"菜单中选择"新建"|"项目"，打开"新建项目"对话框。
(3) 使用模板"006 提取左因子"新建一个项目。

3.2　阅读实验源代码

1. PickupLeftFactor.h 文件（参见源代码清单 6-1）

此文件主要定义了与文法相关的数据结构，这些数据结构定义了文法的单链表存储形式。其中，Rule 结构体用于定义文法的名称和文法链表；RuleSymbol 结构体用于定义文法产生式中的终结符和非终结符。具体内容可参见表 6-1 和表 6-2。

表　6-1

Rule 的域	说　　明
RuleName	文法的名称
pFirstSymbol	指向文法的第一个 Select 的第一个 Symbol
pNextRule	指向下一条文法

表 6-2

RuleSymbol 的域	说 明
pNextSymbol	指向下一个 Symbol
pOther	指向下一个 Select
isToken	是否为终结符。1 表示终结符,0 表示非终结符
TokenName	终结符的名称。isToken 为 1 时这个域有效
pRule	指向 Symbol 对应的 Rule。isToken 为 0 时这个域有效

下面是一个简单文法,并使用图 6-1 和图 6-2 说明了该文法的存储结构:

```
A->Aa|aB
B->bB
```

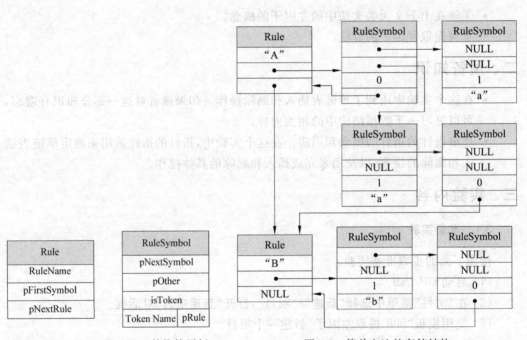

图 6-1 Rule 和 RuleSymbol 结构体图例   图 6-2 简单文法的存储结构

2. main.c 文件(参见源代码清单 6-2)

此文件定义了 main 函数。在 main 函数中首先调用 InitRules 函数初始化了文法,然后调用 PrintRule 函数打印提取左因子之前的文法,接着调用 PickupLeftFactor 函数对文法提取左因子,最后再次调用 PrintRule 函数打印提取左因子之后的文法。

在 main 函数的后面定义了一系列函数,有关这些函数的具体内容参见表 6-3。关于这些函数的参数和返回值,可以参见其注释。

表 6-3

函 数 名	功 能 说 明
GetSymbol	根据下标找到 Select 中的一个 Symbol
LeftFactorMaxLength	以一个 Select 为模板,确定左因子的最大长度 提示: • 虽然 Select 中的 Symbol 是个单链表,但是可以调用 GetSymbol 函数通过下标访问 Symbol,这样可以简化逐个访问 Symbol 的过程 • 当需要比较两个 Symbol 是否相同时,可以调用 SymbolCmp 函数 此函数的函数体还不完整,留给读者完成
SymbolCmp	比较两个相同类型(同为终结符或同为非终结符)的 Symbol 是否具有相同的名字 此函数的函数体还不完整,留给读者完成
NeedPickup	取文法中的一个 Select 与 SelectTemplate 进行比较,判断该 Select 是否需要提取左因子 提示: • 可以调用 GetSymbol 函数通过下标访问 Symbol,这样可以简化逐个访问 Symbol 的过程 • 当需要比较两个 Symbol 是否相同时,可以调用 SymbolCmp 函数 此函数的函数体还不完整,留给读者完成
AddSelectToRule	将一个 Select 加入到文法末尾,当 Select 为 NULL 时就将一个 ε 终结符加入到文法末尾。在本程序中 ε 可以用 $ 来代替 函数的函数体还不完整,留给读者完成
GetUniqueRuleName	此函数会在新文法的名称后面添加后缀,直到新文法的名称唯一
FreeSelect	释放一个 Select 的内存 此函数的函数体还不完整,留给读者完成
PickupLeftFactor	对文法提取左因子 提示: • 由于对一条文法提取左因子之后还可能存在左因子,所以设置了一个标识 isChange,并初始化为 0,每当提取了左因子之后就将该标识赋值为 1,并将此标识作为循环条件,直到没有提取左因子操作发生时,才能结束循环 • 在本函数中使用指针的指针 pSelectPtr 来确定符号在单链表中插入和删除的位置,在进入下一次循环之前应为 pSelectPtr 设置正确的值 此函数的函数体还不完整,留给读者完成
InitRules	使用给定的数据初始化文法链表
CreateRule	创建一个新的 Rule
CreateSymbol	创建一个新的 Symbol
FindRule	根据 RuleName 在文法链表中查找名字相同的文法
PrintRule	输出文法。此函数的函数体还不完整,留给读者完成

源代码清单 6-1: PickupLeftFactor.h 文件

```c
#ifndef _PICKUPLEFTFACTOR_H_
#define _PICKUPLEFTFACTOR_H_

//
//在此处包含 C 标准库头文件
//

#include<stdio.h>

//
//在此处包含其他头文件
//


//
//在此处定义数据结构
//

#define MAX_STR_LENGTH 64

struct _Rule;
typedef struct _RuleSymbol{
    struct _RuleSymbol * pNextSymbol;      //指向下一个 Symbol
    struct _RuleSymbol * pOther;           //指向下一个 Select
    int isToken;                           //是否为终结符。1表示终结符,0表示非终结符
    char TokenName[MAX_STR_LENGTH];        //终结符的名称。isToken 为 1 时这个域有效
    struct _Rule * pRule;
                                           //指向 Symbol 对应的 Rule。isToken 为 0 时这个域有效
}RuleSymbol;

typedef struct _Rule{
    char RuleName[MAX_STR_LENGTH];         //文法的名称
    struct _RuleSymbol * pFirstSymbol;     //指向文法的第一个 Select 的第一个 Symbol
    struct _Rule * pNextRule;              //指向下一条文法
}Rule;

//
//在此处声明函数
//

RuleSymbol * GetSymbol(RuleSymbol * pSelect, int index);
int LeftFactorMaxLength(RuleSymbol * pSelectTemplate);
int SymbolCmp(RuleSymbol * pRuleSymbol1, RuleSymbol * pRuleSymbol2);
```

```c
int NeedPickup(RuleSymbol * pSelectTemplate, int Count, RuleSymbol * pSelect);
void AddSelectToRule(Rule * pRule, RuleSymbol * pRuleSymbol);
void GetUniqueRuleName(Rule * pHead, char * pRuleName);
void PickupLeftFactor(Rule * pHead);
void FreeSelect(RuleSymbol * pSelect);

Rule * InitRules();
Rule * CreateRule(const char * pRuleName);
RuleSymbol * CreateSymbol();
Rule * FindRule(Rule * pHead, const char * RuleName);
void PrintRule(Rule * pHead);

//
//在此处声明全局变量
//

extern const char * VoidSymbol;
extern const char * Postfix;


#endif /* _PICKUPLEFTFACTOR_H_ */
```

源代码清单 6-2: main.c 文件

```c
#include "PickupLeftFactor.h"

const char * VoidSymbol="$ "; //"ε"
const char * Postfix="'";

int main(int argc, char * argv[])
{
    //
    //调用 InitRules 函数初始化文法
    //
    Rule * pHead=InitRules();

    //
    //输出提取左因子之前的文法
    //
    printf("Before Pickup Left Factor:\n");
    PrintRule(pHead);

    //
    //调用 PickupLeftFactor 函数对文法提取左因子
    //
    PickupLeftFactor(pHead);
```

```
        //
            //输出提取左因子之后的文法
            //
            printf("\nAfter Pickup Left Factor:\n");
            PrintRule(pHead);

            return 0;
        }

        /*
        功能:
            根据下标找到 Select 中的一个 Symbol。

        参数:
            pSelect--Select 指针。
            index--下标。

        返回值:
            如果存在,返回找到的 Symbol 指针,否则返回 NULL。
        */
        RuleSymbol * GetSymbol(RuleSymbol * pSelect, int index)
        {
            int i=0;
            RuleSymbol * pRuleSymbol;
            for(pRuleSymbol=pSelect, i=0; pRuleSymbol!=NULL;
                pRuleSymbol=pRuleSymbol->pNextSymbol, i++)
            {
                if(i==index)
                    return pRuleSymbol;
            }

            return NULL;
        }

        /*
        功能:
            以 SelectTemplate 为模板,确定左因子的最大长度。

        参数:
            pSelectTemplate--作为模板的 Select 指针。

        返回值:
```

左因子的最大长度,如果返回 0,说明不存在左因子。
*/
int LeftFactorMaxLength(RuleSymbol * pSelectTemplate)
{

    //
    //TODO:在此处添加代码
    //

}

/*
功能：
    比较两个相同类型(同为终结符或同为非终结符)的 Symbol 是否具有相同的名字。

参数：
    pSymbol1--Symbol 指针。
    pSymbol2--Symbol 指针。

返回值：
    相同返回 1,不同返回 0。
*/
int SymbolCmp(RuleSymbol * pSymbol1, RuleSymbol * pSymbol2)
{

    //
    //TODO:在此处添加代码
    //

}

/*
功能：
    取文法中的一个 Select 与 SelectTemplate 进行比较,判断该 Select 是否需要提取左因子。

参数：
    pSelectTemplate--作为模板的 Select 指针。
    Count--SelectTemplate 中已确定的左因子的数量。
    pSelect--Select 指针。

返回值：
    如果 Select 包含左因子返回 1,否则返回 0。

```c
 */
int NeedPickup(RuleSymbol * pSelectTemplate, int Count, RuleSymbol * pSelect)
{

    //
    //TODO：在此处添加代码
    //


}

/*
功能：
    将一个 Select 加入到文法末尾，当 Select 为 NULL 时就将一个 ε 终结符加入到文法
末尾。

参数：
    pRule--文法指针。
    pNewSelect--Select 指针。
*/
void AddSelectToRule(Rule * pRule, RuleSymbol * pNewSelect)
{

    //
    //TODO：在此处添加代码
    //


}

/*
功能：
    将 pRuleName 与文法中的其他 RuleName 比较，如果相同就增加一个后缀。

参数：
    pHead--Rule 链表的头指针。
    pRuleName--Rule 的名字。
*/
void GetUniqueRuleName(Rule * pHead, char * pRuleName)
{
    Rule * pRuleCursor=pHead;
    for(; pRuleCursor!=NULL;)
    {
        if(0==strcmp(pRuleCursor->RuleName, pRuleName))
        {
```

```
                strcat(pRuleName, Postfix);
                pRuleCursor=pHead;
                continue;
            }
            pRuleCursor=pRuleCursor->pNextRule;
        }
}

/*
功能：
    释放一个 Select 的内存。

参数：
    pSelect--需要释放的 Select 的指针。
*/
void FreeSelect(RuleSymbol * pSelect)
{

    //
    //TODO：在此处添加代码
    //

}

/*
功能：
    提取左因子。

参数：
    pHead--文法的头指针。
*/
void PickupLeftFactor(Rule * pHead)
{
    Rule * pRule;                    //Rule 游标
    int isChange;                    //Rule 是否被提取左因子的标志
    RuleSymbol * pSelectTemplate;    //Select 游标
    Rule * pNewRule;                 //Rule 指针
    RuleSymbol * pSelect;            //Select 游标

    //
    //TODO：在此处添加代码
    //

}
```

```c
/*
功能:
    使用给定的数据初始化文法链表。

返回值:
    文法的头指针。
*/
typedef struct _SYMBOL{
    int isToken;
    char Name[MAX_STR_LENGTH];
}SYMBOL;

typedef struct _RULE_ENTRY{
    char RuleName[MAX_STR_LENGTH];
    SYMBOL Selects[64][64];
}RULE_ENTRY;

static const RULE_ENTRY rule_table[]=
{
    /* A->abC | abcD | abcE */
    { "A",{
            { { 1, "a" }, { 1, "b" }, { 1, "C" } },
            { { 1, "a" }, { 1, "b" }, { 1, "c" }, { 1, "D" } },
            { { 1, "a" }, { 1, "b" }, { 1, "c" }, { 1, "E" } }
        }
    }
};

Rule * InitRules()
{
    Rule * pHead, * pRule;
    RuleSymbol * * pSymbolPtr1, * * pSymbolPtr2;
    int nRuleCount=sizeof(rule_table) / sizeof(rule_table[0]);
    int i,j,k;

    Rule * * pRulePtr=&pHead;
    for(i=0; i<nRuleCount; i++)
    {
        * pRulePtr=CreateRule(rule_table[i].RuleName);
        pRulePtr=&(* pRulePtr)->pNextRule;
    }

    pRule=pHead;
```

```c
    for(i=0; i<nRuleCount; i++)
    {
        pSymbolPtr1=&pRule->pFirstSymbol;
        for(j=0; rule_table[i].Selects[j][0].Name[0]!='\0'; j++)
        {
            pSymbolPtr2=pSymbolPtr1;
            for(k=0; rule_table[i].Selects[j][k].Name[0]!='\0'; k++)
            {
                const SYMBOL * pSymbol=&rule_table[i].Selects[j][k];

                *pSymbolPtr2=CreateSymbol();
                (*pSymbolPtr2)->isToken=pSymbol->isToken;
                if(1==pSymbol->isToken)
                {
                    strcpy((*pSymbolPtr2)->TokenName, pSymbol->Name);
                }
                else
                {
                    (*pSymbolPtr2)->pRule=FindRule(pHead, pSymbol->Name);
                    if(NULL==(*pSymbolPtr2)->pRule)
                    {
                        printf("Init rules error, miss rule \"%s\"\n", pSymbol->Name);
                        exit(1);
                    }
                }
                pSymbolPtr2=&(*pSymbolPtr2)->pNextSymbol;
            }

            pSymbolPtr1=&(*pSymbolPtr1)->pOther;
        }

        pRule=pRule->pNextRule;
    }

    return pHead;
}

/*
功能:
    创建一个新的Rule。

参数:
```

pRuleName--文法的名字。

返回值:
　　Rule 指针。
*/
```
Rule * CreateRule(const char * pRuleName)
{
    Rule * pRule=(Rule *)malloc(sizeof(Rule));

    strcpy(pRule->RuleName, pRuleName);
    pRule->pFirstSymbol=NULL;
    pRule->pNextRule=NULL;

    return pRule;
}
```

/*

功能:
　　创建一个新的 Symbol。

返回值:
　　RuleSymbol 指针。
*/
```
RuleSymbol * CreateSymbol()
{
    RuleSymbol * pSymbol=(RuleSymbol *)malloc(sizeof(RuleSymbol));

    pSymbol->pNextSymbol=NULL;
    pSymbol->pOther=NULL;
    pSymbol->isToken=-1;
    pSymbol->TokenName[0]='\0';
    pSymbol->pRule=NULL;

    return pSymbol;
}
```

/*

功能:
　　根据 RuleName 在文法链表中查找名字相同的文法。

参数:
　　pHead--文法的头指针。
　　RuleName--文法的名字。

```
返回值:
    Rule 指针。
*/
Rule * FindRule(Rule * pHead, const char * RuleName)
{
    Rule * pRule;
    for(pRule=pHead; pRule!=NULL; pRule=pRule->pNextRule)
    {
        if(0==strcmp(pRule->RuleName, RuleName))
        {
            break;
        }
    }

    return pRule;
}

/*
功能:
    输出文法。
参数:
    pHead--文法的头指针。
*/
void PrintRule(Rule * pHead)
{

    //
    //TODO:在此处添加代码
    //

}
```

3.3 为函数 InitRules 添加注释

在 InitRules 函数之前定义了两个结构体,这两个结构体用来定义初始化文法数据的存储形式。具体内容可参见表 6-4 和表 6-5。

表 6-4

SYMBOL 的域	说明
isToken	是否为终结符。1 表示终结符,0 表示非终结符
Name	终结符和非终结符的名称

实验 6 提取左因子

表 6-5

RULE_ENTRY 的域	说 明
RuleName	文法的名称。
Selects	SYMBOL 结构体的二维数组,其中每一行表示一个 Select,一行中的每个元素分别表示一个终结符或非终结符。

为 InitRules 函数添加注释(注意,在本程序中使用了指针的指针,体会其在单链表的插入和删除操作中的作用)。

3.4 为 PrintRule 函数编写源代码

为 PrintRule 函数编写源代码,同时理解在本程序中文法的链式存储结构,编写完源代码后,选择"调试"菜单中的"调试/开始执行(不调试)",会在控制台窗口输出文法的产生式,如图 6-3 所示。由于还没有为 PickupLeftFactor 函数和其他未完成的函数编写源代码,所以前后两次输出的文法是一样的。

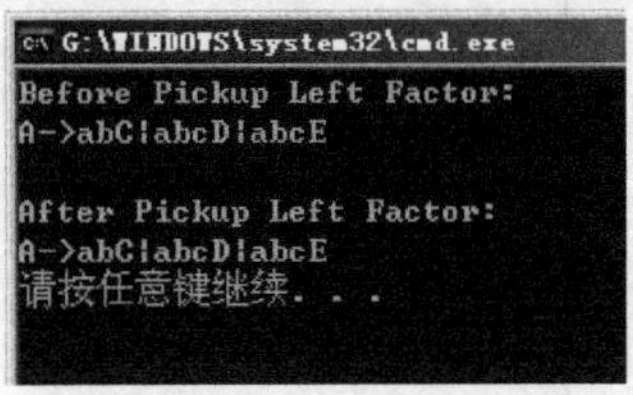

图 6-3 打印文法

3.5 在演示模式下调试项目

按照下面的步骤调试项目:

(1) 按 F7 键生成项目。

(2) 在演示模式下,按 F5 键启动调试项目。程序会在观察点函数的开始位置中断。

(3) 重复按 F5 键,直到调试过程结束。

在调试的过程中,每执行"演示流程"窗口中的一行后,仔细观察"转储信息"窗口内容所发生的变化,理解对文法提取左因子的过程。"转储信息"窗口显示的数据信息包括:

- 文法。在提取左因子的过程中会使用游标来指向正在操作的文法。
- 新文法。新生成的包含原文法左因子之后部分的文法。
- 执行到循环的末尾时,会将新文法加入到文法链表中。
- 在查找包含左因子的 Select 模板过程中会使用空心中括号表示出游标指向的 Select。
- 在提取左因子的过程中会使用加粗的中括号表示出游标指向的 Select。

3.6 编写源代码并通过验证

按照下面的步骤继续实验:

(1) 为 PickupLeftFactor 函数和其他未完成的函数编写源代码。注意尽量使用已定义的局部变量。

(2) 按 F7 键生成项目。如果生成失败,根据"输出"窗口中的提示信息修改源代码中的语法错误。

(3) 按 Alt+F5 键启动验证。如果验证失败,可以使用"输出"窗口中的"比较"功能,或者在"非演示模式"下按 F5 键启动调试后重复按 F10 键单步调试所编写的源代码,从而定位错误的位置,然后回到步骤(1)。

四、思考与练习

1. 请读者仔细检查自己编写的源代码,查看在消除左递归的过程中是否调用了 free 函数释放了从文法链表中移除的 Symbol,是否调用了 FreeSelect 函数释放了从文法链表中移除的 Select,否则会造成内存泄露。

2. 编写一个 FreeRule 函数,当在 main 函数的最后调用此函数时,可以将整个文法的内存释放掉,从而避免内存泄露。

3. 使用自己编写的代码对下面的例子进行验证,确保程序可以为所有形式的文法提取左因子,并验证通过。

```
A->B|abcD|abB|abcE|abcM|D|aB
N->ab|abD
```

4. 本实验使用的例子第一次提取了左因子"ab",请读者尝试第一次提取左因子"abc",并与之前的结果进行比较。然后尝试修改提取左因子的算法,总是从文法中优先提取最长的左因子。

EXPERIMENT 7

实验 7　First 集合

实验难度：★★☆☆☆
建议学时：2 学时

一、实验目的

- 了解在上下文无关文法中的 First 集合的定义。
- 掌握计算 First 集合的方法。

二、预备知识

对 First 集合的定义有初步的理解。读者可以参考配套的《编译原理》教材，预习这部分内容。

三、实验内容

3.1　准备实验

按照下面的步骤准备实验：
(1) 启动 CP Lab。
(2) 在"文件"菜单中选择"新建"|"项目"，打开"新建项目"对话框。
(3) 使用模板"007 First 集合"新建一个项目。

3.2　阅读实验源代码

1. First.h 文件(参见源代码清单 7-1)

此文件主要定义了与文法和集合相关的数据结构。

有关文法的数据结构定义了文法的单链表存储形式。其中，Rule 结构体用于定义文法的名称和文法链表；RuleSymbol 结构体用于定义文法产生式中的终结符和非终结符。(注意：此处的 RuleSymbol 结构体与之前在消除左递归和提取左因子中的 RuleSymbol 结构体有一些差异，原因是在计算 First 集合时，将文法中的每个选择都单独写成了一个产生式，所以 RuleSymbol 结构体就进行了相应的简化。)

有关集合的数据结构定义了集合的线性表存储形式。其中，Set 结构体用于定义集合的名称和终结符数组；SetList 结构体用于定义一个以 Set 为元素的线性表。具体内容可参见表 7-1～表 7-4。

表 7-1

Rule 的域	说　　明
RuleName	文法的名称
pFirstSymbol	指向文法的第一个 Symbol
pNextRule	指向下一条文法

表 7-2

RuleSymbol 的域	说　　明
pNextSymbol	指向下一个 Symbol
isToken	是否为终结符。1 表示终结符,0 表示非终结符
SymbolName	终结符和非终结符的名称

表 7-3

Set 的域	说　　明
Nane	集合的名称
Terminal	终结符数组
nTerminalCount	终结符数组元素个数。与前一项构成一个线性表

表 7-4

SetList 的域	说　　明
Sets	集合数组
nSetCount	集合数组元素个数。与前一项构成一个线性表

下面是一个简单文法,并使用图 7-1 和图 7-2 说明了该文法的存储结构:

A->Aa
A->aB
B->bB

Rule
RuleName
pFirstSymbol
pNextRule

RuleSymbol
pNextSymbol
isToken
SymbolName

图 7-1　Rule 和 RuleSymbol 结构体图例

2. main.c 文件(参见源代码清单 7-2)

此文件定义了 main 函数。在 main 函数中首先调用 InitRules 函数初始化文法,然后调用 PrintRule 函数打印文法,接着初始化保存 First 集合的线性表,最后调用 First 函数求文法的 First 集合。

在 main 函数的后面定义了一系列函数,有关这些函数的具体内容参见表 7-5。关于这些函数的参数和返回值,可以参见其注释。

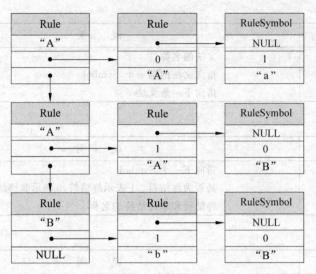

图 7-2 简单文法的存储结构

表 7-5

函 数 名	功 能 说 明
AddOneSet	添加一个 Set 到 SetList 中 提示：在这个函数中可以首先调用一次 GetSet 函数，用来判断该 SetList 是否已经存在同名的 Set，从而忽略重复的名称 此函数的函数体还不完整，留给读者完成
GetSet	根据名称在 SetList 中查找 Set 此函数的函数体还不完整，留给读者完成
AddTerminalToSet	添加一个终结符到 Set 提示：需要忽略重复的终结符 此函数的函数体还不完整，留给读者完成
AddSetToSet	将源 Set 中的所有终结符添加到目标 Set 中 提示： • 需要忽略源 Set 中的 ε • 可以通过调用 AddTerminalToSet 函数将终结符加入到目标 Set 中 此函数的函数体还不完整，留给读者完成
SetHasVoid	判断 Set 的终结符中是否含有 ε 此函数的函数体还不完整，留给读者完成
First	求文法的 First 集合 提示： • 由于对一个文法符号（一个终结符或非终结符）求 First 集合可能会依赖于其他的文法符号的 First 集合，所以设置了一个标识 isChange，并初始化为 0，每当集合列表中的某个集合发生变化后就将该标识赋值为 1，并将此标识作为循环条件，直到没有变化发生时，才能结束循环 • 在循环末尾，如果当前文法产生式的每一个符号的 First 集合都包含 ε，就将 ε 终结符添加到当前文法名称的 First 集合中 此函数的函数体还不完整，留给读者完成

续表

函 数 名	功 能 说 明
InitRules	使用给定的数据初始化文法链表
CreateRule	创建一个新的 Rule
CreateSymbol	创建一个新的 Symbol
PrintRule	输出文法 此函数的函数体还不完整,留给读者完成

源代码清单 7-1：First.h 文件

```c
#ifndef _FIRST_H_
#define _FIRST_H_

//
//在此处包含 C 标准库头文件
//

#include<stdio.h>

//
//在此处包含其他头文件
//

//
//在此处定义数据结构
//

#define MAX_STR_LENGTH 64

struct _Rule;
typedef struct _RuleSymbol{
    struct _RuleSymbol * pNextSymbol;     //指向下一个 Symbol
    int isToken;                          //是否为终结符。1 表示终结符,0 表示非终结符
    char SymbolName[MAX_STR_LENGTH];      //终结符和非终结符的名称
}RuleSymbol;

typedef struct _Rule{
    char RuleName[MAX_STR_LENGTH];        //文法的名称
    struct _RuleSymbol * pFirstSymbol;    //指向文法的第一个 Symbol
    struct _Rule * pNextRule;             //指向下一条文法
}Rule;

typedef struct _Set{
```

实验 7　First 集合

```
    char Name[MAX_STR_LENGTH];              //集合的名称
    char Terminal[32][MAX_STR_LENGTH];      //终结符数组
    int nTerminalCount;                     //数组元素个数
}Set;

typedef struct _SetList{
    Set Sets[32];                           //集合数组
    int nSetCount;                          //数组元素个数
}SetList;

//
//在此处声明函数
//

void First(const Rule * pHead, SetList * pFirstSetList);
Set * GetSet(SetList * pSetList, const char * pName);
void AddOneSet(SetList * pSetList, const char * pName);
int AddTerminalToSet(Set * pSet, const char * pTerminal);
int AddSetToSet(Set * pDesSet, const Set * pSrcSet);
int SetHasVoid(const Set * pSet);

Rule * InitRules();
Rule * CreateRule(const char * pRuleName);
RuleSymbol * CreateSymbol();

void PrintRule(const Rule * pHead);

//
//在此声明全局变量
//

extern const char * VoidSymbol;

#endif /* _FIRST_H_ */
```

源代码清单 7-2：main.c 文件

```
#include "First.h"

const char * VoidSymbol="$ ";               //"ε"

int main(int argc, char * argv[])
{
    //
```

```
    //调用InitRules函数初始化文法
    //
    Rule * pHead=InitRules();

    //
    //输出文法
    //
    PrintRule(pHead);

    //
    //调用First函数求文法的First集合
    //
    SetList FirstSet;
    FirstSet.nSetCount=0;
    First(pHead, &FirstSet);

    return 0;
}

/*
功能：
    添加一个Set到SetList。

参数：
    pSetList--SetList指针。
    pName--集合名称字符串指针。
*/
void AddOneSet(SetList * pSetList, const char * pName)
{

    //
    //TODO：在此添加代码
    //

}

/*
功能：
    根据名称在SetList中查找Set。

参数：
    pSetList--SetList指针。
    pName--集合名称字符串指针。
```

返回值：
　　如果找到返回 Set 的指针，否则返回 NULL。
*/
Set * GetSet(SetList * pSetList, const char * pName)
{

　　//
　　//TODO：在此处添加代码
　　//

}

/*
功能：
　　添加一个终结符到 Set。

参数：
　　pSet--Set 指针。
　　pTerminal--终结符名称指针。

返回值：
　　添加成功返回 1。
　　添加不成功返回 0。
*/
int AddTerminalToSet(Set * pSet, const char * pTerminal)
{

　　//
　　//TODO：在此处添加代码
　　//

}

/*
功能：
　　将源 Set 中的所有终结符添加到目标 Set 中。

参数：
　　pDesSet--目标 Set 指针。
　　pSrcSet--源 Set 指针。

返回值：
　　添加成功返回 1，否则返回 0。

```
*/
int AddSetToSet(Set * pDesSet, const Set * pSrcSet)
{

    //
    //TODO: 在此处添加代码
    //

}

/*
功能:
    判断 Set 的终结符中是否含有 ε。

参数:
    pSet--Set 指针。

返回值:
    存在返回 1。
    不存在返回 0。
*/
int SetHasVoid(const Set * pSet)
{

    //
    //TODO: 在此处添加代码
    //

}

/*
功能:
    求文法的 First 集合。

参数:
    pHead--文法的头指针。
    pFirstSetList--First 集合指针。
*/
void First(const Rule * pHead, SetList * pFirstSetList)
{
    const Rule * pRule;         //Rule 指针
    int isChange;               //集合是否被修改的标志
    RuleSymbol * pSymbol;       //Symbol 游标
```

```
    //
    //TODO：在此处添加代码
    //

}

typedef struct _SYMBOL{
    int isToken;
    char SymbolName[MAX_STR_LENGTH];
}SYMBOL;

typedef struct _RULE_ENTRY{
    char RuleName[MAX_STR_LENGTH];
    SYMBOL Symbols[64];
}RULE_ENTRY;

static const RULE_ENTRY rule_table[]=
{
    /* exp->exp addop term| term */
    { "exp", { { 0, "exp" }, { 0, "addop"}, { 0, "term"} } },
    { "exp", { { 0, "term" } } },

    /* addop->+|-*/
    { "addop", { { 1, "+" } } },
    { "addop", { { 1, "-" } } },

    /* term->term mulop factor | factor */
    { "term", { { 0, "term" }, { 0, "mulop"}, { 0, "factor"} } },
    { "term", { { 0, "factor" } } },

    /* mulop-> * */
    { "mulop", { { 1, "*" } } },

    /* factor->(exp) | number */
    { "factor", { { 1, "(" }, { 0, "exp"}, { 1, ")" } } },
    { "factor", { { 1, "number" } } },
};

/*
功能：
    初始化文法链表。

返回值：
```

        文法的头指针。
*/
Rule * InitRules()
{
    Rule * pHead, * pRule;
    RuleSymbol * * pSymbolPtr, * pNewSymbol;
    int nRuleCount=sizeof(rule_table) / sizeof(rule_table[0]);
    int i, j;

    Rule * * pRulePtr=&pHead;
    for(i=0; i<nRuleCount; i++)
    {
        * pRulePtr=CreateRule(rule_table[i].RuleName);
        pRulePtr=&(* pRulePtr)->pNextRule;
    }

    pRule=pHead;
    for(i=0; i<nRuleCount; i++)
    {
        pSymbolPtr=&pRule->pFirstSymbol;
        for(j=0; rule_table[i].Symbols[j].SymbolName[0]!='\0'; j++)
        {
            const SYMBOL * pSymbol=&rule_table[i].Symbols[j];

            pNewSymbol=CreateSymbol();
            pNewSymbol->isToken=pSymbol->isToken;
            strcpy(pNewSymbol->SymbolName, pSymbol->SymbolName);
            * pSymbolPtr=pNewSymbol;

            pSymbolPtr=&pNewSymbol->pNextSymbol;
        }

        pRule=pRule->pNextRule;
    }

    return pHead;
}

/*
功能：
    创建一个新的文法。

参数：

pRuleName--文法的名字。

返回值：
    文法的指针。
*/
```
Rule * CreateRule(const char * pRuleName)
{
    Rule * pRule=(Rule * )malloc(sizeof(Rule));

    strcpy(pRule->RuleName, pRuleName);
    pRule->pFirstSymbol=NULL;
    pRule->pNextRule=NULL;

    return pRule;
}
```

/*
功能：
    创建一个新的符号。

返回值：
    符号的指针。
*/
```
RuleSymbol * CreateSymbol()
{
    RuleSymbol * pSymbol=(RuleSymbol * )malloc(sizeof(RuleSymbol));

    pSymbol->pNextSymbol=NULL;
    pSymbol->isToken=-1;
    pSymbol->SymbolName[0]='\0';

    return pSymbol;
}
```

/*
功能：
    输出文法。

参数：
    pHead--文法的头指针。
*/
```
void PrintRule(const Rule * pHead)
{
```

```
        //
        //TODO：在此处添加代码
        //

}
```

3.3 为函数 InitRules 添加注释

在 InitRules 函数之前定义了两个结构体，这两个结构体用来定义初始化文法数据的存储形式。具体内容可参见表 7-6 和表 7-7。

表 7-6

SYMBOL 的域	说 明
isToken	是否为终结符。1 表示终结符，0 表示非终结符
SymbolName	终结符或非终结符的名称

表 7-7

RULE_ENTRY 的域	说 明
RuleName	文法的名称
Symbols	SYMBOL 结构体数组，其中每个元素表示一个终结符或非终结符

为 InitRules 函数添加注释（注意，在本程序中使用了指针的指针，体会其在单链表的插入和删除操作中的作用）。

3.4 为 PrintRule 函数编写源代码

为 PrintRule 函数编写源代码，同时理解在本程序中文法的链式存储结构，编写完源代码后，选择"调试"菜单中的"调试/开始执行（不调试）"，会在控制台窗口输出文法的产生式，如图 7-3 所示。

图 7-3 打印文法

3.5 在演示模式下调试项目

按照下面的步骤调试项目：

（1）按 F7 键生成项目。

（2）在演示模式下，按 F5 键启动调试项目。程序会在观察点函数的开始位置中断。

（3）重复按 F5 键，直到调试过程结束。

在调试的过程中，每执行"演示流程"窗口中的一行后，仔细观察"转储信息"窗口内容所发生的变化，理解求文法的 First 集合的过程。"转储信息"窗口显示的数据信息包括：

- 在求文法的 First 集合的过程中会使用游标来指向正在操作的文法。
- 在求文法的 First 集合的过程中会使用加粗的中括号表示出游标指向的 Symbol。

- First 集合列表。显示每一个文法符号的 First 集合的构造过程。

3.6 编写源代码并通过验证

按照下面的步骤继续实验：

（1）为 First 函数和其他未完成的函数编写源代码。注意尽量使用已定义的局部变量。

（2）按 F7 键生成项目。如果生成失败，根据"输出"窗口中的提示信息修改源代码中的语法错误。

（3）按 Alt＋F5 键启动验证。如果验证失败，可以使用"输出"窗口中的"比较"功能，或者在"非演示模式"下按 F5 键启动调试后重复按 F10 键单步调试所编写的源代码，从而定位错误的位置，然后回到步骤（1）。

四、思考与练习

1. 编写一个 FreeRule 函数，当在 main 函数的最后调用此函数时，可以将整个文法的内存释放掉，从而避免内存泄露。

2. 使用自己编写的代码对下面的例子进行验证，确保程序可以为所有形式的文法求 First 集合，并验证通过。（文法中的终结符用粗体表示。）

A->B
A->ε
B->C
B->ε
C->AB
C-> *

EXPERIMENT 8

实验 8　Follow 集合

实验难度：★★☆☆☆
建议学时：2 学时

一、实验目的

- 了解在上下文无关文法中的 First 集合和 Follow 集合的定义。
- 掌握计算 First 集合和 Follow 集合的方法。

二、预备知识

对 First 集合和 Follow 集合的定义有初步的理解。读者可以参考配套的《编译原理》教材，预习这一部分内容。

三、实验内容

3.1　准备实验

按照下面的步骤准备实验：
(1) 启动 CP Lab。
(2) 在"文件"菜单中选择"新建"|"项目"，打开"新建项目"对话框。
(3) 使用模板"008 Follow 集合"新建一个项目。

3.2　阅读实验源代码

1. Follow.h 文件（参见源代码清单 8-1）

此文件主要定义了与文法和集合相关的数据结构。

有关文法的数据结构定义了文法的单链表存储形式。其中 Rule 结构体用于定义文法的名称和文法链表；RuleSymbol 结构体用于定义文法产生式中的终结符和非终结符。（**注意**：此处的 RuleSymbol 结构体与之前在消除左递归和提取左因子中的 RuleSymbol 结构体有一些差异，原因是在计算 First 集合和 Follow 集合时，将文法中的每个选择都单独写成了一个产生式，所以 RuleSymbol 结构体就进行了相应的简化。）

有关集合的数据结构定义了集合的线性表存储形式。其中，Set 结构体用于定义集合的名称和终结符数组；SetList 结构体用于定义一个以 Set 为元素的线性表。具体内容可参见表 8-1 至表 8-4。

表 8-1

Rule 的域	说明
RuleName	文法的名称
pFirstSymbol	指向文法的第一个 Symbol
pNextRule	指向下一条文法

表 8-2

RuleSymbol 的域	说明
pNextSymbol	指向下一个 Symbol
isToken	是否为终结符。1 表示终结符,0 表示非终结符
SymbolName	终结符和非终结符的名称

表 8-3

Set 的域	说明
Nane	集合的名称
Terminal	终结符数组
nTerminalCount	终结符数组元素个数。与前一项构成一个线性表

表 8-4

SetList 的域	说明
Sets	集合数组
nSetCount	集合数组元素个数。与前一项构成一个线性表

下面是一个简单文法,并使用图 8-1 和图 8-2 说明了该文法的存储结构:

A->Aa
A->aB
B->bB

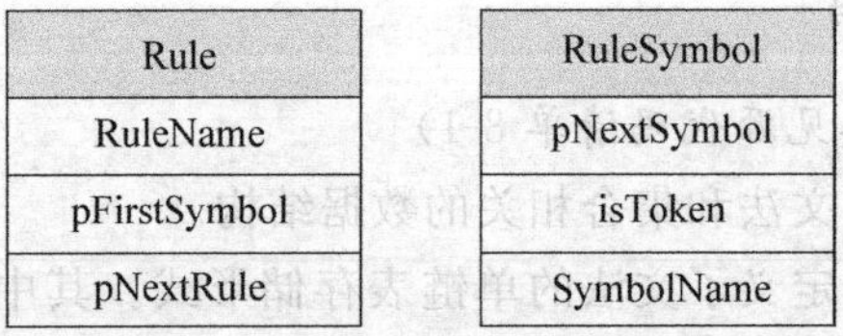

Rule
RuleName
pFirstSymbol
pNextRule

RuleSymbol
pNextSymbol
isToken
SymbolName

图 8-1 Rule 和 RuleSymbol 结构体图例

2. main.c 文件(参见源代码清单 8-2)

此文件定义了 main 函数。在 main 函数中首先调用 InitRules 函数初始化了文法,然后调用 PrintRule 函数打印文法,接着初始化了保存 First 集合和 Follow 集合的线性表,最后调用 Follow 函数求文法的 First 集合和 Follow 集合。

在 main 函数的后面定义了一系列函数,有关这些函数的具体内容参见表 8-5。关于

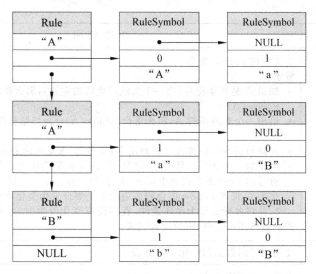

图 8-2 简单文法的存储结构

这些函数的参数和返回值,可以参见其注释。

表 8-5

函 数 名	功 能 说 明
AddOneSet	添加一个 Set 到 SetList 中 提示:在这个函数中可以首先调用一次 GetSet 函数,用来判断该 SetList 是否已经存在同名的 Set,从而忽略重复的名称 此函数的函数体还不完整,留给读者完成
GetSet	根据名称在 SetList 中查找 Set 此函数的函数体还不完整,留给读者完成
AddTerminalToSet	添加一个终结符到 Set 提示:需要忽略重复的终结符 此函数的函数体还不完整,留给读者完成
AddSetToSet	将源 Set 中的所有终结符添加到目标 Set 中 提示: • 需要忽略源 Set 中的 ε • 可以通过调用 AddTerminalToSet 函数将终结符加入到目标 Set 中。此函数的函数体还不完整,留给读者完成。
SetHasVoid	判断 Set 的终结符中是否含有 ε 此函数的函数体还不完整,留给读者完成
First	求文法的 First 集合 提示: • 由于对一个文法符号(一个终结符或非终结符)求 First 集合可能会依赖于其他的文法符号的 First 集合,所以设置了一个标识 isChange,并初始化为 0,每当集合列表中的某个集合发生变化后就将该标识赋值为 1,并将此标识作为循环条件,直到没有变化发生时,才能结束循环 • 在循环末尾,如果当前文法产生式的每一个符号的 First 集合都包含 ε,就将 ε 终结符添加到当前文法名称的 First 集合中。 此函数的函数体还不完整,留给读者完成

续表

函 数 名	功 能 说 明
Follow	求文法的 Follow 集合 提示： • 如果 A 是开始符号(第一个文法产生式的左边)，那么就将终结符"＄"加入 Follow(A) • 给出一个非终结符 A，那么集合 Follow(A)是由终结符组成，此外可能还有"＄" • 在函数的开始位置首先要调用 First 函数，求文法的 First 集合 • 由于对一个文法符号(非终结符)求 Follow 集合可能会依赖于其他的文法符号的 First 集合或者 Follow 集合，所以设置了一个标识 isChange，并初始化为 0，每当 Follow 集合列表中的某个集合发生变化后就将该标识赋值为 1，并将此标识作为循环条件，直到没有变化发生时，才能结束循环此函数的函数体还不完整，留给读者完成
InitRules	使用给定的数据初始化文法链表
CreateRule	创建一个新的 Rule
CreateSymbol	创建一个新的 Symbol
PrintRule	输出文法 此函数的函数体还不完整，留给读者完成

源代码清单 8-1：Follow.h 文件

```
#ifndef _FOLLOW_H_
#define _FOLLOW_H_

//
//在此处包含 C 标准库头文件
//

#include<stdio.h>


//
//在此处包含其他头文件
//


//
//在此处定义数据结构
//

#define MAX_STR_LENGTH 64

struct _Rule;
typedef struct _RuleSymbol{
    struct _RuleSymbol * pNextSymbol;       //指向下一个 Symbol
    int isToken;                            //是否为终结符。1表示终结符，0表示非终结符
```

编译原理实验教程

```c
    char SymbolName[MAX_STR_LENGTH];    //终结符和非终结符的名称
}RuleSymbol;

typedef struct _Rule{
    char RuleName[MAX_STR_LENGTH];      //文法的名称
    struct _RuleSymbol * pFirstSymbol;  //指向文法的第一个Symbol
    struct _Rule * pNextRule;           //指向下一条文法
}Rule;

typedef struct _Set{
    char Name[MAX_STR_LENGTH];          //集合的名称
    char Terminal[32][MAX_STR_LENGTH];  //终结符数组
    int nTerminalCount;                 //数组元素个数
}Set;

typedef struct _SetList{
    Set Sets[32];                       //集合数组
    int nSetCount;                      //数组元素个数
}SetList;

//
//在此处声明函数
//

void First(const Rule * pHead, SetList * pFirstSetList);
void Follow (const Rule * pHead, SetList * pFollowSetList, SetList * pFirstSetList);
Set * GetSet(SetList * pSetList, const char * pName);
void AddOneSet(SetList * pSetList, const char * pName);
int AddTerminalToSet(Set * pSet, const char * pTerminal);
int AddSetToSet(Set * pDesSet, const Set * pSrcSet);
int SetHasVoid(const Set * pSet);

Rule * InitRules();
Rule * CreateRule(const char * pRuleName);
RuleSymbol * CreateSymbol();

void PrintRule(const Rule * pHead);

//
//在此处声明全局变量
```

```
//
extern const char * VoidSymbol;
const char * DollarSymbol;

#endif /* _FOLLOW_H_ */
```

源代码清单 8-2：main.c 文件

```c
#include "Follow.h"

const char * VoidSymbol="#";           //"ε"
const char * DollarSymbol="$ ";

int main(int argc, char * argv[])
{
    //
    //调用 InitRules 函数初始化文法
    //
    Rule * pHead=InitRules();

    //
    //初始化 First 集合、Follow 集合
    //
    SetList FirstSetList, FollowSetList;
    FirstSetList.nSetCount=0;
    FollowSetList.nSetCount=0;

    //
    //调用 Follow 函数求文法的 First 集合、Follow 集合
    //
    Follow(pHead, &FollowSetList, &FirstSetList);

    //
    //输出文法
    //
    PrintRule(pHead);

    return 0;
}
```

```
/*
功能：
    添加一个 Set 到 SetList。

参数：
    pSetList--SetList 指针。
    pName--集合名称字符串指针。
*/
void AddOneSet(SetList * pSetList, const char * pName)
{

    //
    //TODO：在此处添加代码
    //

}

/*
功能：
    根据名称在 SetList 中查找。

参数：
    pSetList--SetList 指针。
    pName--集合名称字符串指针。

返回值：
    如果找到返回 Set 指针,否则返回 NULL。
*/
Set * GetSet(SetList * pSetList, const char * pName)
{

    //
    //TODO：在此处添加代码
    //

}

/*
功能：
    添加一个终结符到 Set。

参数：
    pSet--Set 指针。
    pTerminal--终结符名称指针。
```

返回值：
    添加成功返回 1，否则返回 0。
*/
int AddTerminalToSet(Set * pSet, const char * pTerminal)
{

    //
    //TODO：在此处添加代码
    //

}

/*
功能：
    将源 Set 添加到目标 Set 中，忽略 ε。

参数：
    pDesSet--目标 Set 指针。
    pSrcSet--源 Set 指针。

返回值：
    添加成功返回 1，否则返回 0。
*/
int AddSetToSet(Set * pDesSet, const Set * pSrcSet)
{

    //
    //TODO：在此处添加代码
    //

}

/*
功能：
    判断 Set 中是否含有 ε。

参数：
    pSet--Set 指针。

返回值：
    存在返回 1。
    不存在返回 0。
*/

```
int SetHasVoid(const Set * pSet)
{

    //
    //TODO:在此处添加代码
    //

}
```

/*
功能：
    求文法的 First 集合。

参数：
    pHead--文法的头指针。
    pFirstSetList--First 集合指针。
*/
```
void First(const Rule * pHead, SetList * pFirstSetList)
{
    const Rule * pRule;         //Rule 指针
    int isChange;               //集合是否被修改的标志
    RuleSymbol * pSymbol;       //Symbol 游标

    //
    //TODO:在此处添加代码
    //

}
```

/*
功能：
    求文法的 Follow 集合。

参数：
    pHead--文法的头指针。
    pFollowSetList--Follow 集合指针。
    pFirstSetList--First 集合指针。
*/
```
void Follow (const Rule * pHead, SetList * pFollowSetList, SetList * pFirstSetList)
{
    const Rule * pRule;         //Rule 指针
    int isChange;               //集合是否被修改的标志
```

```c
    RuleSymbol * pSymbol;              //Symbol 游标

    //调用 First 函数求文法的 First 集合
    First(pHead, pFirstSetList);

    //
    //TODO: 在此处添加代码
    //

}

typedef struct _SYMBOL{
    int isToken;
    char SymbolName[MAX_STR_LENGTH];
}SYMBOL;

typedef struct _RULE_ENTRY{
    char RuleName[MAX_STR_LENGTH];
    SYMBOL Symbols[64];
}RULE_ENTRY;

static const RULE_ENTRY rule_table[]=
{
    /* exp->exp addop term| term */
    { "exp", { { 0, "exp" }, { 0, "addop"}, { 0, "term"} } },
    { "exp", { { 0, "term" } } },

    /* addop->+|- */
    { "addop", { { 1, "+" } } },
    { "addop", { { 1, "-" } } },

    /* term->term mulop factor | factor */
    { "term", { { 0, "term" }, { 0, "mulop"}, { 0, "factor"} } },
    { "term", { { 0, "factor" } } },

    /* mulop-> * */
    { "mulop", { { 1, "*" } } },

    /* factor->(exp) | number */
    { "factor", { { 1, "(" }, { 0, "exp"}, { 1, ")"} } },
    { "factor", { { 1, "number" } } },
};
```

```
/*
功能：
    初始化文法链表。

返回值：
    文法的头指针。
*/
Rule * InitRules()
{
    Rule * pHead, * pRule;
    RuleSymbol * * pSymbolPtr, * pNewSymbol;
    int nRuleCount=sizeof(rule_table) / sizeof(rule_table[0]);
    int i, j;

    Rule * * pRulePtr=&pHead;
    for(i=0; i<nRuleCount; i++)
    {
        * pRulePtr=CreateRule(rule_table[i].RuleName);
        pRulePtr=&(* pRulePtr)->pNextRule;
    }

    pRule=pHead;
    for(i=0; i<nRuleCount; i++)
    {
        pSymbolPtr=&pRule->pFirstSymbol;
        for(j=0; rule_table[i].Symbols[j].SymbolName[0]!='\0'; j++)
        {
            const SYMBOL * pSymbol=&rule_table[i].Symbols[j];

            pNewSymbol=CreateSymbol();
            pNewSymbol->isToken=pSymbol->isToken;
            strcpy(pNewSymbol->SymbolName, pSymbol->SymbolName);
            * pSymbolPtr=pNewSymbol;

            pSymbolPtr=&pNewSymbol->pNextSymbol;
        }

        pRule=pRule->pNextRule;
    }

    return pHead;
}
```

```
/*
功能：
    创建一个新的文法。

参数：
    pRuleName--文法的名字。

返回值：
    文法的指针。
*/
Rule * CreateRule(const char * pRuleName)
{
    Rule * pRule=(Rule * )malloc(sizeof(Rule));

    strcpy(pRule->RuleName, pRuleName);
    pRule->pFirstSymbol=NULL;
    pRule->pNextRule=NULL;

    return pRule;
}

/*
功能：
    创建一个新的符号。

返回值：
    符号的指针。
*/
RuleSymbol * CreateSymbol()
{
    RuleSymbol * pSymbol=(RuleSymbol * )malloc(sizeof(RuleSymbol));

    pSymbol->pNextSymbol=NULL;
    pSymbol->isToken=-1;
    pSymbol->SymbolName[0]='\0';

    return pSymbol;
}

/*
功能：
    输出文法。
```

```
参数：
    pHead--文法的头指针。
*/
void PrintRule(const Rule * pHead)
{
    const Rule * pRule;
    for(pRule=pHead; pRule!=NULL; pRule=pRule->pNextRule)
    {
        printf("%s->", pRule->RuleName);

        RuleSymbol * pRuleSymbol;
        for(pRuleSymbol=pRule->pFirstSymbol; pRuleSymbol!=NULL;
            pRuleSymbol=pRuleSymbol->pNextSymbol)
        {
            printf(" %s", pRuleSymbol->SymbolName);
        }
        printf("\n");
    }
}
```

3.3 为函数 InitRules 添加注释

在 InitRules 函数之前定义了两个结构体，这两个结构体用来定义初始化文法数据的存储形式。具体内容可参见表 8-6 和表 8-7。

表 8-6

SYMBOL 的域	说　　明
isToken	是否为终结符。1 表示终结符，0 表示非终结符
SymbolName	终结符和非终结符的名称

表 8-7

RULE_ENTRY 的域	说　　明
RuleName	文法的名称
Symbols	SYMBOL 结构体数组，其中每个元素表示一个终结符或非终结符

为 InitRules 函数添加注释。（注意，在本程序中使用了指针的指针，体会其在单链表的插入和删除操作中的作用。）

3.4 为 PrintRule 函数编写源代码

为 PrintRule 函数编写源代码，同时理解在本程序中文法的链式存储结构，编写完源代码后，选择"调试"菜单中的"调试/开始执行（不调试）"，会在控制台窗口输出文法的产生式，如图 8-3 所示。

图 8-3 打印文法

3.5 在演示模式下调试项目

按照下面的步骤调试项目：

(1) 按 F7 键生成项目。

(2) 在演示模式下，按 F5 键启动调试项目。程序会在观察点函数的开始位置中断。

(3) 重复按 F5 键，直到调试过程结束。

在调试的过程中，每执行"演示流程"窗口中的一行后，仔细观察"转储信息"窗口内容所发生的变化，理解求文法的 Follow 集合的过程。"转储信息"窗口显示的数据信息包括：

- 在求文法的 Follow 集合的过程中使用游标来指向正在操作的文法。
- 在求文法的 Follow 集合的过程中使用加粗的中括号表示出游标指向的 Symbol。
- First 集合列表。
- Follow 集合列表。显示每一个文法符号的 Follow 集合的构造过程。

3.6 编写源代码并通过验证

按照下面的步骤继续实验：

(1) 为 Follow 函数和其他未完成的函数编写源代码。注意尽量使用已定义的局部变量。

(2) 按 F7 键生成项目。如果生成失败，根据"输出"窗口中的提示信息修改源代码中的语法错误。

(3) 按 Alt+F5 键启动验证。如果验证失败，可以使用"输出"窗口中的"比较"功能，或者在"非演示模式"下按 F5 键启动调试后重复按 F10 键单步调试所编写的源代码，从而定位错误的位置，然后回到步骤(1)。

四、思考与练习

1. 编写一个 FreeRule 函数，当在 main 函数的最后调用此函数时，可以将整个文法的内存释放掉，从而避免内存泄露。

2. 使用自己编写的代码对下面的例子进行验证，确保程序可以为所有形式的文法求 Follow 集合，并验证通过（文法中的终结符用粗体表示）。

```
A->B
A->ε
B->C
B->ε
C->AB
C->*
```

3. 编写一个程序,此程序可以根据 First 集合和 Follow 集合判断文法是否为 LL(1) 文法。

4. 编写一个程序,此程序可以根据 LL(1) 文法的 First 集合和 Follow 集合构造 LL(1) 分析表。

5. 编写一个程序,此程序可以使用 LL(1) 分析表自动分析输入的字符串,并输出 LL(1) 的分析动作和错误信息。

EXPERIMENT 9

实验 9 Yacc 分析程序生成器

实验难度：★★☆☆☆
建议学时：2 学时

一、实验目的

- 掌握 Yacc 输入文件的格式。
- 掌握使用 Yacc 自动生成分析程序的方法。

二、预备知识

- 要求已经学习了 BNF(上下文无关文法)，能够正确编写简单的 BNF。
- 熟练掌握了各种自底向上的分析方法，特别是 Yacc 所使用的 LALR(1)算法。
- 本实验使用 Yacc 的一个实现版本——GNU Bison 作为分析程序生成器。

三、实验内容

3.1 准备实验

按照下面的步骤准备实验：
(1) 启动 CP Lab。
(2) 在"文件"菜单中选择"新建"|"项目"，打开"新建项目"对话框。
(3) 使用模板"009 Yacc 分析程序生成器"新建一个项目。

3.2 阅读实验源代码

1. sample.txt 文件(参见源代码清单 9-1)

此文件是 Yacc 的输入文件。根据 Yacc 输入文件的格式，此文件分为 3 个部分(由%%分隔)，各个部分的说明可以参见表 9-1。

表 9-1

名称	说明
第一部分	在%{和%}之间直接插入 C 源代码文件的内容。包括要包含的头文件，以及用于指示分析程序输出调试信息的 YYDEBUG 预定义类型
第二部分	定义了一个简单的文法规则"A—>(A)|a"，但是，对规则进行匹配时，没有执行任何操作
第三部分	这部分中的内容会直接插入 C 源代码文件。这部分内容的说明可以参见表 9-2

表 9-2

内容	说明
main 函数	在 main 函数中首先声明了 Yacc 的全局变量 yydebug,然后将其赋值为 0(0 表示不启动调试),最后调用 yyparse 函数对从标准输入读取到的内容进行分析
yylex 函数	此函数从标准输入读取字符并进行相关的处理。首先会忽略包括空格在内的空白,然后,如果输入了回车,就返回 0,让分析程序停止,否则就返回从标准输入读取到的字符,用于和文法进行匹配
yyerror 函数	当从标准输入读取到的字符与文法匹配失败时,会调用此函数输出错误信息

2. ytab.c 文件

此文件是 Yacc 输出的 C 源代码文件。当使用 Yacc 处理 sample.txt 文件时,就会生成此文件。新建项目中,此文件的内容是空的。

3. ytab.h 文件

此文件是 Yacc 输出的头文件。为 Yacc 使用选项"--defines=ytab.h"时,就会生成此文件。此文件可以被包括在需要使用 Yacc 所生成的定义的任何文件中。新建项目中,此文件的内容是空的。

4. y.output.txt 文件

此文件是 Yacc 输出的文件。为 Yacc 使用选项"--report-file=y.output.txt"时,就会生成此文件。此文件包含了被分析程序使用的 LALR(1) 分析表的文本描述。新建项目中,此文件的内容是空的。

5. y.output.html 文件

此文件是 y.output.txt 文件内容的 HTML 语言表示,可以使用更加直观的方式显示分析表的信息。新建项目中,此文件的内容是空的。

6. y.dot.txt 文件

此文件是 y.output.txt 文件内容的 DOT 语言表示,可以使用图形化的方式显示 DFA 自动机。新建项目中,此文件的内容是空的。

源代码清单 9-1: sample.txt 文件

```
%{

#include<stdio.h>
#include<ctype.h>

#define YYDEBUG 1

%}

%%

command: A
       ;
```

```
/* A->(A)|a */
A : 'a'
  | '(' A ')'
  ;

%%

main()
{
    extern int yydebug;
    yydebug=0;

    return yyparse();
}

int yylex(void)
{
    int c;
    while((c=getchar())==' ');        /*忽略空格*/

    if(c=='\n') return 0;             /*停止*/

    return c;
}

/*打印错误信息*/
int yyerror(char * s)
{
    fprintf(stdout, "%s\n", s);
    return 0;
}
```

3.3 生成项目

按照下面的步骤生成项目：

(1) 在"生成"菜单中选择"重新生成项目"(快捷键是 Ctrl+Alt+F7)。

(2) 在生成的过程中，CP Lab 会首先使用 Bison 程序根据输入文件 sample.txt 来生成各个输出文件，然后，将生成的 ytab.c 文件重新编译、连接为可以运行的可执行文件。

(3) 如果成功生成了 ytab.c 文件，CP Lab 还会自动使用 DOTTY 程序来打开 y.dot.txt 文件，读者可以使用图形化的方式查看 DFA 自动机。

(4) 读者可以在"项目管理器"窗口中双击 y.output.html 文件，使用浏览器打开此文件,其内容与 y.output.txt 文件类似,但是查看更加方便直观(注意,如果浏览器中显示乱码,需要将浏览器的编码改为"UTF-8",简单的修改方法是在乱码页面中右击,选择"编码"中的"UTF-8")。

(5) 在生成的 ytab.c 文件中,尝试找到 sample.txt 文件中第一部分和第三部分 C 源代码插入的位置,并尝试查找 yylex 函数和 yyerror 函数是在哪里被调用的。

提示:如果需要使用 DOTTY 程序手动打开 y.dot.txt 文件,需要首先在 CP Lab 的"工具"菜单中选择 Dotty,然后在 DOTTY 程序中右击,选择菜单中的 load graph,打开项目目录中的 y.dot.txt 文件。

注意:在下面的实验步骤中,如果需要生成项目,应尽量使用"重新生成项目"功能。如果习惯使用"生成项目"(快捷键是 F7)功能,可能需要连续使用两次此功能才能生成最新的项目。

3.4 运行项目

在没有对项目的源代码进行任何修改的情况下,按照下面的步骤运行项目:
(1) 选择"调试"菜单中的"开始执行(不调试)"(快捷键是 Ctrl+F5)。
(2) 在 Windows 控制台窗口中输入"(a)"字符串后按回车,扫描程序不会输出任何错误信息,如图 9-1 所示,说明文法匹配成功。而如果在 Windows 控制台窗口中输入类似"()"或"b"字符串后按回车,就会输出默认的错误信息,如图 9-2 所示。

图 9-1 字符串匹配成功

图 9-2 字符串匹配失败

3.5 编写一个简单的计算器程序

按照下面的步骤完成此练习:
(1) 修改 sample.txt 文件中的内容,实现一个简单的计算器程序,其文法如下(粗体表示终结符):

exp→exp addop term | term
addop→**+** | **-**
term→term mulop factor | factor
mulop→*****
factor→**(** exp **)** | number

(2) 重新生成项目。如果生成失败,根据"输出"窗口中的提示信息修改源代码中的语法错误。
(3) 按 Ctrl+F5 键启动执行项目。

执行的结果如图 9-3 所示,当在 Windows 控制台窗口中输入表达式"2+3"后按回车,可以正确计算出表达式的结果。

图 9-3　正确计算表达式的结果

注意:

(1) 本实验的模板不提供演示功能,所以,如果执行的结果不正确,可以通过添加断点和单步调试的方法来查找错误的原因。

(2) 断点应该添加在 sample.txt 文件中需要中断的 C 源代码行,不要添加在 ytab.c 文件中,否则无法命中断点。

四、思考与练习

1. 尝试为计算器文法绘制 LALR(1) 的分析表,并绘制表达式"2+3"的分析动作表。提示:将 main 函数中的 yydebug 赋值为 1 后,就可以在 Windows 控制台窗口中获得分析程序的分析动作。

2. 尝试为计算器程序添加整除运算符"/",并可以为包含整除运算符的表达式计算出正确的值。

3. 修改计算器的 YACC 输入文件,使之能够输出以下有用的错误信息:

- 为表达式"(2 +3"生成错误信息"丢失右括号";
- 为表达式"2+3)"生成错误信息"丢失左括号";
- 为表达式"2 3"生成错误信息"丢失运算符";
- 为表达式"(2+)"生成错误信息"丢失操作数"。

4. 选择 CP Lab"帮助"菜单中"其他帮助文档"中的"Bison 手册",学习 Bison 工具的更多用法。如果需要修改 Bison 程序在处理输入文件时的选项,可以在"项目管理器"窗口中右击文件 sample.txt,在弹出的快捷菜单中选择"属性",然后选择"属性页"左侧"自定义生成步骤"的"常规",就可以编辑右侧的"命令行"选项了。

EXPERIMENT 10

实验 10　符号表的构建和使用

实验难度：★★☆☆☆
建议学时：2 学时

一、实验目的

- 了解符号表的结构。
- 掌握符号表的插入、查找和删除等基本操作。

二、预备知识

- 学习了符号表的概念以及作用域的基本规则。
- 在这个实验中主要用到了杂凑表（哈希表）的插入和查找等操作。如果读者对这一部分知识有遗忘，可以复习一下数据结构中的相关内容。

三、实验内容

3.1　准备实验

按照下面的步骤准备实验：

（1）启动 CP Lab。
（2）在"文件"菜单中选择"新建"|"项目"，打开"新建项目"对话框。
（3）使用模板"010 符号表的构建和使用"新建一个项目。

3.2　阅读实验源代码

1. SymbolTable.h 文件（参见源代码清单 10-1）

此文件主要定义了与符号表相关的数据结构，这些数据结构定义了一个符号表的单链表存储形式，允许每个作用域使用独立的符号表。其中，Symbol 结构体用于定义符号的相关信息和 Symbol 单链表；SymbolTable 结构体用于定义作用域对应的杂凑表和 SymbolTable 单链表。具体内容可参见表 10-1 和表 10-2。

表 10-1

Symbol 的域	说　　明
SymbolName	符号名称
SymbolType	符号类型
ClashCount	冲突次数。当一个符号名称在其作用域中被重复定义时，此计数器就要增加 1

续表

Symbol 的域	说明
RefCount	引用次数。当一个符号名称在其作用域中被引用时,此计数器就要增加 1
pNext	指向下一个 Symbol

表 10-2

SymbolTable 的域	说明
Bucket	杂凑表(桶)
Invalid	作用域是否无效的标志。1 表示无效,0 表示有效
pNext	指向下一个 SymbolTable

2. main.c 文件(参见源代码清单 10-2)

首先定义了符号表链表的头指针和符号引用失败计数器两个全局变量,然后定义了 main 函数。在 main 函数中调用了 CreateSymbolTable 函数构造符号表。

在 main 函数的后面定义了一系列函数,有关这些函数的具体内容参见表 10-3。关于这些函数的参数和返回值,可以参见其注释。

表 10-3

函 数 名	功 能 说 明
NewSymbol	创建一个新的符号 注意,调用了 memset 函数将符号的内容清空,这与将符号结构体中的每个字段分别清空是等效的,但是使用 memset 函数可以编写更少的代码,执行效率更高
NewSymbolTable	创建一个新的符号表 注意,也调用了 memset 函数将符号表的内容清空
PushScope	在符号表链表的表头添加一个作用域 提示: • 可以调用 NewSymbolTable 函数创建符号表 • 添加一个作用域和入栈操作类似 此函数的函数体还不完整,留给读者完成
PopScope	将符号表链表中最内层的作用域设置为无效 提示: • 并不需要将符号表从符号表链表中移除,只需将该符号表中的 Invalid 域置为 1 即可 • 设置一个作用域无效和出栈操作类似 此函数的函数体还不完整,留给读者完成
AddSymbol	向符号表中添加一个 Symbol 提示: • 每当定义一个新的变量时,会调用此函数向符号表中添加一个 Symbol • 如果新定义变量的名称在其作用域中与已定义变量的名称重复,就不能添加 Symbol,而应将已定义变量的 ClashCount 域增加 1 此函数的函数体还不完整,留给读者完成

续表

函 数 名	功 能 说 明
RefSymbol	对 Symbol 进行一次引用 提示： • 每当使用一个已定义的变量时，会调用此函数将所用变量的 RefCount 域增加 1 • 如果使用的变量未定义，此函数需将符号引用失败计数器(RefErrorCount)增加 1 此函数的函数体还不完整，留给读者完成
Hush	求 Symbol 的哈希值

CreateSymbolTable	提示：
	• 通过调用 PushScope 函数模拟进入作用域
	• 通过调用 PopScope 函数模拟退出作用域
	• 通过调用 AddSymbol 函数模拟定义一个变量
	• 通过调用 RefSymbol 函数模拟使用一个变量
	此函数中的代码模拟了下列源代码在进入作用域、退出作用域、定义变量和使用变量时对符号表的操作
	<pre>{
	    int i, j; int f(); //函数声明
	    {
	        char i;
	        int size;
	        char temp;
	        {
	            char * j;
	            long j; //重复定义
	            j=NULL;
	            i='p';
	            size=f();
	            new=0; //引用失败
	        }
	        j=3;
	    }
	}</pre>

源代码清单 10-1: SymbolTable.h 文件

```
#ifndef _SYMBOL_TABLE_H_
#define _SYMBOL_TABLE_H_

//
//在此处包含 C 标准库头文件
//

#include<stdio.h>
//
```

实验 10　符号表的构建和使用

```c
//在此处包含其他头文件
//

//
//在此处定义数据结构
//
#define MAX_STR_LENGTH      64
#define BUCKET_SIZE         5               //选择一个素数作为桶大小

typedef struct _Symbol{
    char SymbolName[MAX_STR_LENGTH];        //符号名称
    char SymbolType[MAX_STR_LENGTH];        //符号类型
    int ClashCount;                         //冲突次数
    int RefCount;                           //引用次数
    struct _Symbol * pNext;                 //指向下一个Symbol
}Symbol;

typedef struct _SymbolTable{
    Symbol * Bucket[BUCKET_SIZE];           //杂凑表(桶)
    int Invalid;                            //作用域是否无效的标志。1表示无效,0表示有效
    struct _SymbolTable * pNext;            //指向下一个SymbolTable
}SymboTable;

//
//在此处声明函数
//

Symbol * NewSymbol();
SymboTable * NewSymbolTable();
void PushScope();
void PopScope();
void RefSymbol(const char * pSymbolName);
void AddSymbol(const char * pSymbolName, const char * pSymbolType);
int Hush(const char * pSymbolName);
void CreateSymbolTable();

//
//在此处声明全局变量
//

extern SymboTable * pHead;
extern int RefErrorCount;

#endif                                      //_SYMBOL_TABLE_H_
```

源代码清单 10-2：main.c 文件

```c
#include "SymbolTable.h"

SymboTable * pHead=NULL;            //符号表链表的头指针
int RefErrorCount=0;                //符号引用失败计数器

int main()
{
    //
    //构造符号表
    //
    CreateSymbolTable();

    return 0;
}


/*
功能：
    创建一个新的符号。

返回值：
    符号的指针。
*/
Symbol * NewSymbol()
{
    Symbol * pNewSymbol= (Symbol * )malloc(sizeof(Symbol));
    memset(pNewSymbol, 0, sizeof(Symbol));

    return pNewSymbol;
}


/*
功能：
    创建一个新的符号表。

返回值：
    符号表的指针。
*/
SymboTable * NewSymbolTable()
{
    SymboTable * pNewTable= (SymboTable * )malloc(sizeof(SymboTable));
    memset(pNewTable, 0, sizeof(SymboTable));
```

```
        return pNewTable;
}

/*
功能:
    在符号表链表的表头添加一个作用域。
*/
void PushScope()
{

    //
    //TODO: 在此添加代码
    //

}

/*
功能:
    将符号表链表中最内层的作用域设置为无效。
*/
void PopScope()
{

    //
    //TODO: 在此添加代码
    //

}

/*
功能:
    对 Symbol 进行一次引用。

参数:
    pSymbolName--符号名称字符串指针。
*/
void RefSymbol(const char * pSymbolName)
{

    //
    //TODO: 在此添加代码
    //

}
```

```
/*
功能：
    向符号表中添加一个 Symbol。

参数：
    pSymbolName--符号名称字符串指针。
    pSymbolType--符号类型字符串指针。
*/
void AddSymbol(const char * pSymbolName, const char * pSymbolType)
{

    //
    //TODO：在此添加代码
    //

}

/*
功能：
    求 Symbol 的哈希值。

参数：
    pSymbolName--符号名称字符串指针。
*/
int Hush(const char * pSymbolName)
{
    int HashValue=0, i;

    for(i=0; pSymbolName[i]!='\0'; i++)
    {
        HashValue=((HashValue<<4)+pSymbolName[i])%BUCKET_SIZE;
    }

    return HashValue;
}

/*
功能：
    创建符号表。
*/
void CreateSymbolTable()
{
    PushScope();
    AddSymbol("i", "int");
    AddSymbol("j", "int");
```

```
        AddSymbol("f", "funciton");

        {
            PushScope();
            AddSymbol("i", "char");
            AddSymbol("size", "int");
            AddSymbol("temp", "char");

            {
                PushScope();
                AddSymbol("j", "char * ");
                AddSymbol("j", "long");

                RefSymbol("j");
                RefSymbol("i");
                RefSymbol("f");
                RefSymbol("size");
                RefSymbol("new");
                PopScope();
            }

            RefSymbol("j");
            PopScope();
        }

        PopScope();
    }
```

3.3 在演示模式下调试项目

按照下面的步骤调试项目:
(1) 按 F7 键生成项目。
(2) 在演示模式下,按 F5 键启动调试项目。程序会在观察点函数的开始位置中断。
(3) 重复按 F5 键,直到调试过程结束。

在调试的过程中,每执行"演示流程"窗口中的一行后,仔细观察"转储信息"窗口内容所发生的变化,理解符号表的构造和使用过程。"转储信息"窗口显示的数据信息包括:
- 符号引用失败次数。
- 作用域是否无效。
- 符号表链表。包括索引、项链表以及每一项的引用次数和冲突次数。
- 显示每一个符号表的构造过程。处在相同作用域的不重名变量,如果哈希值相同,则将新定义的变量插入项列表的表头。

3.4 编写源代码并通过验证

按照下面的步骤继续实验：

（1）为函数体不完整的函数编写源代码。注意尽量使用已定义的局部变量。

（2）按 F7 键生成项目。如果生成失败,根据"输出"窗口中的提示信息修改源代码中的语法错误。

（3）按 Alt+F5 键启动验证。如果验证失败,可以使用"输出"窗口中的"比较"功能,或者在"非演示模式"下按 F5 键启动调试后重复按 F10 键单步调试所编写的源代码,从而定位错误的位置,然后回到步骤(1)。

提示：如果需要通过验证,请不要修改 CreateSymbolTable 函数中的代码,只需将其他函数体补充完整即可。

四、思考与练习

1. 编写一个 FreeSymbolTable 函数,当在 main 函数的最后调用此函数时,可以将整个符号表链表的内存释放掉,从而避免内存泄露。

2. 编写一个 ScanSymbolTable 函数,当在 main 函数的最后调用此函数时,可以对整个符号表进行扫描,对于未被引用的变量发出警告。

3. 修改 PopScope 函数,不再使用符号表的无效标志,而是将作用域对应的符号表从链表头移除,并考虑此时如何对未被引用的变量发出警告。

EXPERIMENT 11

实验 11 三地址码转换为 P-代码

实验难度：★★☆☆☆
建议学时：2 学时

一、实验目的

- 了解三地址码和 P-代码的定义。
- 实现三地址码到 P-代码的转换。

二、预备知识

- 对三地址码和 P-代码的定义有初步的理解，了解三地址码的四元式实现的数据结构。读者可以参考配套的《编译原理》教材，预习这一部分内容。
- 了解三地址码指令和 P-代码指令的用法和对应关系。

三、实验内容

3.1 准备实验

按照下面的步骤准备实验：
(1) 启动 CP Lab。
(2) 在"文件"菜单中选择"新建"|"项目"，打开"新建项目"对话框。
(3) 使用模板"011 三地址码转换为 P-代码"新建一个项目。

3.2 阅读实验源代码

1. T2P.h 文件（参见源代码清单 11-1）

此文件主要定义了与三地址码和 P-代码相关的数据结构。

三地址码使用了四元式的数据结构。其中，AddrKind 是枚举类型，表示地址的类型可以是空、整数常量或字符串，Address 结构体用于定义三地址码中的地址，包括地址类型以及地址中的具体值，TOpKind 是枚举类型，表示三地址码中的指令类型，TCode 结构体用于定义一个指令类型和 3 个地址。

P-代码的数据结构也包括地址的数据结构，与三地址码不同的是 P-代码结构体中只包含一个地址。另外，P-代码拥有自己的指令集。具体内容可参见表 11-1 至表 11-3。

表 11-1

Address 的域	说 明
Kind	地址类型。这个域是一个枚举类型
Value	地址的整数值。地址类型为整数常量时，这个域有效
Name	地址的字符串。地址类型为字符串时，这个域有效

表 11-2

TCode 的域	说 明
Kind	三地址码指令类型。这个域是一个枚举类型
Addr1,Addr2,Addr3	3 个地址

表 11-3

PCode 的域	说 明
Kind	P-代码指令类型。这个域是一个枚举类型
Addr	地址

2. main.c 文件(参见源代码清单 11-2)

此文件定义了 main 函数。在 main 函数中首先定义了 TCodeList 和 PCodeList 两个数组，然后调用了 memset 函数将两个数组的内容清空(使用 memset 函数可以编写更少的代码，执行效率更高)，接着调用 InitTCodeList 函数初始化三地址码数组，最后调用 T2P 函数将三地址码转换为 P-代码。

在 main 函数的后面定义了一系列函数，有关这些函数的具体内容参见表 11-4。关于这些函数的参数和返回值，可以参见其注释。

表 11-4

函 数 名	功 能 说 明
T2P	将三地址码转换为 P-代码。 注意：在函数中使用了 switch 语句，根据三地址码的指令类型来转换为 P-代码，所以需要了解每条三地址码指令和 P-代码指令的用途和对应关系 此函数的函数体还不完整，留给读者完成。
InitAddress	使用给定的数据初始化三地址码。 注意：在初始化的过程中，由于初始化结构体中统一用字符串来存储地址的值，所以当地址类型为 intconst 时需要调用 atoi 函数转换为整型
InitCodeList	初始化三地址码列表。在此函数中会调用 InitAddress 函数

在 InitAddress 函数之前定义了两个结构体，这两个结构体用来定义初始化三地址码的存储形式。具体内容可参见表 11-5 和表 11-6。

表 11-5

AddressEntry 的域	说 明
Kind	地址的类型。这个域是一个枚举类型
Content	地址的值

表 11-6

TCodeEntry 的域	说 明
Kind	三地址码指令类型。这个域是一个枚举类型
Addr1,Addr2,Addr3	3 个地址

源代码清单 11-1：T2P.h 文件

```c
#ifndef _T2P_H_
#define _T2P_H_

//
//在此处包含C标准库头文件
//

#include<stdio.h>

//
//在此处包含其他头文件
//

//
//在此处定义数据结构
//

#define MAX_STR_LENGTH      64
#define MAX_CODE_COUNT      64

//地址类型
typedef enum _AddrKind{
    empty,                          //空
    intconst,                       //整数常量
    string                          //字符串
}AddrKind;

                                    //地址
typedef struct _Address{
    AddrKind Kind;                  //地址的类型。这个域是一个枚举类型
    int Value;                      //地址的值。地址类型为整数常量时,这个域有效
    char Name[MAX_STR_LENGTH];      //地址的值。地址类型为字符串时,这个域有效
}Address;

//三地址码指令类型
```

```
typedef enum _TOpKind{
    t_rd=1,                          //注意,从1开始
    t_gt,
    t_if_f,
    t_asn,
    t_lab,
    t_mul,
    t_sub,
    t_eq,
    t_wri,
    t_halt
}TOpKind;

//三地址码
typedef struct _TCode{
    TOpKind Kind;                    //三地址码指令类型
    Address Addr1, Addr2, Addr3;     //3个地址
}TCode;

//P-代码指令类型
typedef enum _POpKind{
    p_lda=1,                         //注意,从1开始
    p_rdi,
    p_lod,
    p_ldc,
    p_grt,
    p_fjp,
    p_sto,
    p_lab,
    p_mpi,
    p_sbi,
    p_equ,
    p_wri,
    p_stp
}POpKind;

//P-代码
typedef struct _PCode{
    POpKind Kind;                    //P-代码指令类型
    Address Addr;                    //地址
}PCode;
```

```
//
//在此处声明函数
//

void T2P(TCode * TCodeList, PCode * PCodeList);
void InitTCodeList(TCode * pTCodeList);


//
//在此声明全局变量
//


#endif //_T2P_H_
```

源代码清单 11-2: **main.c** 文件

```c
#include "T2P.h"

int main()
{
    TCode TCodeList[MAX_CODE_COUNT];        //三地址码列表
    PCode PCodeList[MAX_CODE_COUNT];        //P-代码列表

    //将三地址码列表和 P-代码列表的内容清空
    memset(TCodeList,0,sizeof(TCodeList));
    memset(PCodeList,0,sizeof(PCodeList));

    //
    //初始化三地址码列表
    //
    InitTCodeList(TCodeList);

    //
    //将三地址码转换为 P-代码
    //
    T2P(TCodeList, PCodeList);

    return 0;
}

/*
功能:
    将三地址码转换为 P-代码。
```

参数：
    TCodeList--三地址码列表指针。
    PCodeList--P-代码列表指针。
*/
```
void T2P(TCode * TCodeList, PCode * PCodeList)
{
    int TIndex=0;                    //三地址码列表游标
    int PIndex=0;                    //P-代码列表游标

    //
    //TODO: 在此添加代码
    //

}

typedef struct _AddressEntry
{
    AddrKind Kind;
    const char * Content;
}AddressEntry;

typedef struct _TCodeEntry
{
    TOpKind Kind;
    AddressEntry Addr1, Addr2, Addr3;
}TCodeEntry;

static const TCodeEntry TCodeTable[]=
{
    { t_rd, { string, "x" } },
    { t_mul, { string, "x" }, { intconst, "2" }, { string, "t1"} },
    { t_asn, { string, "t1" }, { string, "x" } },
    { t_if_f, { string, "x" }, { string, "L1" } },
    { t_wri, { string, "x" } },
    { t_lab, { string, "L1" } },
    { t_halt }
};
```

/*
功能：
    初始化地址。

参数：

```
    pEntry--用于初始化地址的结构体。
    pAddr--地址指针。
*/
void InitAddress(const AddressEntry * pEntry, Address * pAddr)
{
    pAddr->Kind=pEntry->Kind;
    switch(pAddr->Kind)
    {
    case empty:
        break;
    case intconst:
        pAddr->Value=atoi(pEntry->Content);
        break;
    case string:
        strcpy(pAddr->Name, pEntry->Content);
        break;
    }
}

/*
功能：
    初始化三地址码列表。

参数：
    pTCodeList--三地址码列表指针。
*/
void InitTCodeList(TCode * pTCodeList)
{
    int i;
    int EntryCount=sizeof(TCodeTable) / sizeof(TCodeTable[0]);

    for(i=0; i<EntryCount; i++)
    {
        pTCodeList[i].Kind=TCodeTable[i].Kind;

        InitAddress(&TCodeTable[i].Addr1, &pTCodeList[i].Addr1);
        InitAddress(&TCodeTable[i].Addr2, &pTCodeList[i].Addr2);
        InitAddress(&TCodeTable[i].Addr3, &pTCodeList[i].Addr3);
    }
}
```

3.3 在演示模式下调试项目

在本实验中默认例子的源代码是：

```
read x;
x :=x * 2;
if x then
    write x
end
```

这段源代码对应的三地址码是：

```
read x
t1=x * 2
x=t1
if_false x goto L1
write x
label L1
halt
```

以上三地址码对应的四元式已经写到本实验模板的源代码中。

按照下面的步骤调试项目：

(1) 按 F7 键生成项目。
(2) 在演示模式下，按 F5 键启动调试项目。程序会在观察点函数的开始位置中断。
(3) 重复按 F5 键，直到调试过程结束。

在调试的过程中，每执行"演示流程"窗口中的一行后，仔细观察"转储信息"窗口内容所发生的变化，理解三地址码到 P-代码的转换过程。"转储信息"窗口显示的数据信息包括：

- 三地址码(四元式)数组。使用游标来指向正在进行转换的三地址码。
- P-代码数组。显示每一条 P-代码的产生过程。

3.4 编写源代码并通过验证

按照下面的步骤继续实验：

(1) 为 T2P 函数编写源代码。注意尽量使用已定义的局部变量。
(2) 按 F7 键生成项目。如果生成失败，根据"输出"窗口中的提示信息修改源代码中的语法错误。
(3) 按 Alt＋F5 键启动验证。如果验证失败，可以使用"输出"窗口中的"比较"功能，或者在"非演示模式"下按 F5 键启动调试后重复按 F10 键单步调试所编写的源代码，从而定位错误的位置，然后回到步骤(1)。

3.5 一个更为复杂的例子

源代码：

```
read x;
if 0<x then
    fact :=1;
    repeat
        fact :=fact * x;
        x :=x - 1
    until x=0;
    write fact
end
```

对应的三地址码：

```
read x
t1=x>0
if_false t1 goto L1
fact=1
label L2
t2=fact * x
fact=t2
t3=x-1
x=t3
t4=x==0
if_false t4 goto L2
write fact
label L1
halt
```

三地址码对应的四元式：

```
(rd, x, _, _)
(gt, x, 0, t1)
(if_f, t1, L1, _)
(asn, 1, fact, _)
(lab, L2, _, _)
(mul, fact, x, t2)
(asn, t2, fact, _)
(sub, x, 1, t3)
(asn, t3, x, _)
(eq, x, 0, t4)
(if_f, t4, L2, _)
(wri, fact, _, _)
(lab, L1, _, _)
(halt, _, _, _)
```

读者可以将以上的四元式作为三地址码的初始化数据写到本实验的源代码中，并验证通过。

四、思考与练习

1. 将以下源代码的三地址码转换为 P-代码,并验证通过。

```
i:=5*6;
j:=40-i;
if(20>j && 10==j) then
    write j
end
```

2. 改进现有三地址码转换为 P-代码的程序,在得到的 P-代码中尽量减少临时变量的使用。

3. 编写一个 P-代码到三地址码的转换程序。

EXPERIMENT 12

实验 12　GCC 编译器案例综合研究

实验难度：★★☆☆☆
建议学时：2 学时

一、实验目的

- 了解 GCC 提供的 C 编译器。
- 掌握 GCC 提供的 C 编译器在 32 位 Windows 操作系统上产生的汇编代码，及 C 语言运行时环境。

二、预备知识

- 要求已经学习了编译器生成目标机器的可执行代码的相关知识。
- 学习了基于栈和堆的运行时环境。
- 对本实验使用的 C 编译器案例——GCC 有一定的了解。

三、实验内容

3.1 准备实验

按照下面的步骤准备实验：
(1) 启动 CP Lab。
(2) 在"文件"菜单中选择"新建"|"项目"，打开"新建项目"对话框。
(3) 使用模板"012 GCC 编译器案例综合研究"新建一个项目。

3.2 阅读实验源代码

本实验模板创建的项目中包括两个文件：case.h（参见源代码清单 12-1）和 case.c（参见源代码清单 12-2）。

由于本实验主要关注于编译器产生的汇编代码，使用的 C 代码都非常简单，所以在后面的实验步骤中再对相关的源代码进行介绍。

源代码清单 12-1: case.h

```
#ifndef _CASE_H_
#define _CASE_H_
```

```c
#include<stdio.h>

void case1();
void case2();
void case3();
void case4();
void case5();

int add(int i, int j);

#endif /* _CASE_H_ */
```

源代码清单 12-2: case.c

```c
#include "case.h"

int main(int argc, char * argv[])
{
    case1();
    case2();
    case3();
    case4();
    case5();

    return 0;
}

void case1()
{
    int i=add(1, 2);
}

int add(int i, int j)
{
    return i +j;
}

void case2()
{
    int i=1;
    int a[4];
```

```c
    a[0]=6;
    a[1]=7;
    a[i+1]=8;
}

typedef struct _Rec
{
    int i;
    char c;
    int j;
}Rec;
```

```c
void case3()
{
    Rec x;
    x.j=1;
    x.c=x.j;
    x.i=x.j;
}

typedef struct _TreeNode
{
    int val;
    struct _TreeNode * lchild, * rchild;
}TreeNode;

void case4()
{
    TreeNode * p= (TreeNode * )malloc(sizeof(TreeNode));
    p->lchild=p;
    p=p->rchild;
}

void case5()
{
    int x, y;
    if(x>y)
        y++;
    else
        x--;
}
```

3.3 案例 1——基于栈的函数调用

按照下面的步骤完成此练习：
（1）在 case1 函数唯一的代码行添加一个断点。
（2）按 F7 键生成项目。
（3）按 F5 键启动调试项目。此时程序会在刚刚添加的断点处中断。
（4）选择"调试"菜单"窗口"中的"反汇编"，打开"反汇编"窗口。

在打开的"反汇编"窗口中显示了 case1 函数的反汇编代码，如下面的代码清单所示（为了方便对代码进行说明，添加了行号）：

```
1) push    ebp
2) mov     ebp,esp
3) sub     esp,0x18
4) mov     DWORD PTR [esp+0x4],0x2
5) mov     DWORD PTR [esp],0x1
6) call    0x401359<add>
7) mov     DWORD PTR [ebp-0x4],eax
8) leave
9) ret
```

下面对汇编代码进行详细说明：

（1）32 位 Windows 操作系统一般运行在 32 位的 Intel x86 处理器上，此时使用 32 位的寄存器。为了与 Intel 8086 处理器上 16 位寄存器的名称相区分，在 16 位寄存器名称的前面增加了字母"e"。例如，ax 是 16 位寄存器的名称，相应的 eax 是 32 位寄存器的名称。

（2）第 1 行和第 2 行代码初始化了 case1 函数的栈，如图 12-1 所示，此时 ebp 和 esp 指向相同的地址。由于 ebp 是 32 位寄存器，所以当其压入栈后会占用 4 个字节（深色表示已经压入栈的字节）。注意，这里用到的栈是从高地址向低地址方向生长，esp 总是指向栈顶元素，ebp 总是指向当前函数的"栈底"，esp 和 ebp 之间的栈空间用来存放当前函数的临时变量。

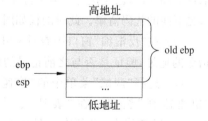

图 12-1　初始化栈

（3）如前所述，第 3 行代码在栈上开辟了一个 24（0x18）字节的空间，用来存放当前函数的临时变量，此时的栈如图 12-2 所示。

（4）第 4 行和第 5 行代码将调用 add 函数所需的两个参数按照从右到左的顺序压入到栈中，如图 12-3 所示。其中，在 esp＋0x4 两边加中括号表示这个地址处的内存，而 DWORD PTR 表示这个内存的大小是 4 字节。由于第 3 行代码已经开辟了栈空间，所以这里直接使用了 mov 指令，请读者练习使用 push 指令改写这两行代码，并适当修改第 3 行代码，使栈增长的空间仍然为 24 个字节。

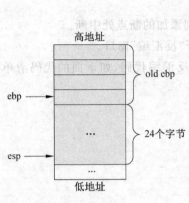

图 12-2　开辟栈空间

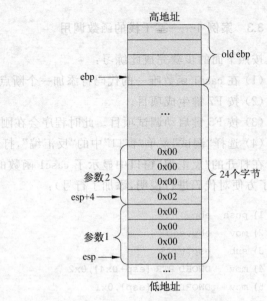

图 12-3　将参数压入栈

(5) 第 6 行代码使用 call 指令调用 add 函数。指令 call 会首先将 add 函数返回后继续执行的地址(也就是第 7 行代码的地址)压入栈，然后跳转到 add 函数第一条指令的地址开始执行。请读者记录下 call 指令跳转的地址以及第 7 行代码所在的地址，留待后面进行验证。

按照下面的步骤继续此练习：

(1) 按 F11 键单步进入 add 函数，会在 add 函数内中断，并在"反汇编"窗口中显示 add 函数的反汇编代码(已经执行了第 1 行和第 2 行代码)，参见下面的代码清单。此时的栈如图 12-4 所示。

(2) 在"反汇编"窗口中查看 add 函数第一条指令的地址，验证是否与之前记录的地址相同。

(3) 选择"调试"菜单中的"快速监视"，打开"快速监视"窗口。在"表达式"控件中输入"＄ebp"后按回车(必须有开始的 ＄ 符号)，记录下此时 ebp 寄存器指向的地址。在"表达式"控件中重新输入"*((long *)(ebp 指向的地址＋4))"即可查看由 call 指令压入栈中的返回地址，验证是否与之前记录的地址相同。

```
1) push    ebp
2) mov     ebp,esp
3) mov     eax,DWORD PTR [ebp+0xc]
4) add     eax,DWORD PTR [ebp+0x8]
```

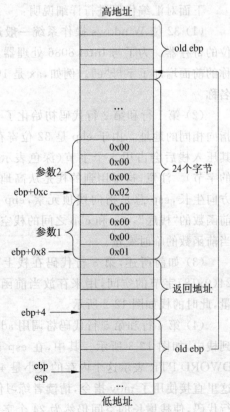

图 12-4　进入 add 函数后的栈

```
5) pop    ebp
6) ret
```

下面继续对汇编代码进行详细说明：

（1）add 函数的第 1 行和第 2 行代码初始化了栈。在第 1 行压入栈中的 ebp 就是之前 case1 函数的"栈底"，在图 12-4 中使用了一个箭头表示这种链接关系；第 2 行将 epb 指向了 add 函数的"栈底"。

（2）第 3 行将参数 2 的值复制到了累加寄存器 eax 中。

（3）第 4 行将参数 1 的值累加到了 eax 中，并且使用 eax 作为函数的返回值。

（4）第 5 行从栈中弹出了 ebp，这会让 ebp 重新回到图 12-3 中所示的位置。

（5）第 6 行的 ret 指令从栈中弹出了函数的返回地址到指令计数器（eip 寄存器）中，这样，整个栈就恢复到如图 12-3 所示的样子了，而且会从 case1 函数反汇编代码的第 7 行继续执行。

（6）case1 函数的第 7 行将 add 函数的返回值从 eax 寄存器复制到临时变量 i 中。由于 int 类型占用 4 个字节，所以，ebp-4 的位置就是临时变量 i 在栈中的位置。请读者在图 12-3 中添加临时变量 i 在栈中的位置。

（7）case1 函数第 8 行的 leave 指令与下面的两条指令

```
mov    esp,ebp
pop    ebp
```

完全等价，其中第 1 条指令负责展开栈，包括释放了传递给 add 函数的参数；第 9 行的 ret 指令会让程序返回到 main 函数中继续执行。

至此，可以对 C 语言函数调用做如下总结：

（1）使用栈传递参数，并按照从右到左的顺序将参数压入栈。

（2）使用 eax 寄存器保存函数的返回值。

（3）由调用者负责展开栈，并释放参数占用的栈空间。

3.4 案例 2——数组元素的引用

按照下面的步骤完成此练习：

（1）按 Shift＋F5 键结束之前的调试。

（2）按 Ctrl＋Shift＋F9 键删除之前添加的所有断点。

（3）在 case2 函数第一行代码处添加一个断点。

（4）按 F5 键启动调试项目。此时程序会在刚刚添加的断点处中断。

（5）选择"调试"菜单"窗口"中的"反汇编"，打开"反汇编"窗口。

在打开的"反汇编"窗口中显示了 case2 函数的反汇编代码，如下面的代码清单所示（为了方便对代码进行说明，添加了行号）：

```
1)  push   ebp
2)  mov    ebp,esp
3)  sub    esp,0x28
```

4) mov   DWORD PTR [ebp-0xc],0x1
5) mov   DWORD PTR [ebp-0x28],0x6
6) mov   DWORD PTR [ebp-0x24],0x7
7) mov   eax,DWORD PTR [ebp-0xc]
8) mov   DWORD PTR [ebp+eax*4-0x24],0x8
9) leave
10) ret

下面对汇编代码进行详细说明：

（1）第 3 行代码在栈上开辟了一个 40(0x28)字节的空间，用来存放当前函数的临时变量，包括变量 i 和数组 a。

（2）第 4 行代码将变量 i 的值赋值为 1。

（3）第 5 行和第 6 行代码分别为 $a[0]$ 和 $a[1]$ 赋值。

（4）第 7 行代码将变量 i 的值复制到 eax 寄存器中。

（5）第 8 行代码使用公式 ebp$-$0x28$+$(eax$+$1)$*$4 计算数组元素的地址并完成赋值操作。

请读者参考图 12-3，绘制出案例 2 中栈的图示，并在栈中标示出临时变量 i 的位置和数组 a 中各个元素的位置。

3.5 案例 3——结构体中域的引用

按照下面的步骤完成此练习：

（1）按 Shift+F5 键结束之前的调试。

（2）按 Ctrl+Shift+F9 键删除之前添加的所有断点。

（3）在 case3 函数第 1 行代码处添加一个断点。

（4）按 F5 键启动调试项目。此时程序会在刚刚添加的断点处中断。

（5）选择"调试"菜单"窗口"中的"反汇编"，打开"反汇编"窗口。

在打开的"反汇编"窗口中显示了 case3 函数的反汇编代码，如下面的代码清单所示（为了方便对代码进行说明，添加了行号）：

1) push   ebp
2) mov   ebp,esp
3) sub   esp,0x18
4) mov   DWORD PTR [ebp-0x10],0x1
5) mov   eax,DWORD PTR [ebp-0x10]
6) mov   BYTE PTR [ebp-0x14],al
7) mov   eax,DWORD PTR [ebp-0x10]
8) mov   DWORD PTR [ebp-0x18],eax
9) leave
10) ret

下面对汇编代码进行详细说明：

（1）第 3 行代码在栈上开辟了一个 24(0x18)字节的空间，用来存放当前函数的临时变量 x。域 x.i 的大小是 4 字节,对应的地址是 ebp-0x18;域 x.c 的大小是 1 字节,按 4 字节对齐后大小也是 4 字节,但是只有最低的一个字节有效,对应的地址是 ebp-0x14;域 x.j 的大小是 4 字节,对应的地址是 ebp-0x10。

(2) 第 4 行代码将 x.j 的值赋值成 1。

(3) 第 5 行和第 6 行代码将 x.j 的值赋值给 x.c。注意,第 6 行代码只是将 eax 最低的一个字节 al 的值赋值给了 x.c 占用的一个字节(由 BYTE PTR 指定为一个字节)。

(4) 第 7 行和第 8 行代码将 x.j 的值赋值给 x.i。

请读者参考图 12-3,绘制出案例 3 中栈的图示,并在栈中标示出临时变量 x 的各个域的位置。

3.6 案例 4——指针的引用

按照下面的步骤完成此练习:

(1) 按 Shift+F5 键结束之前的调试。
(2) 按 Ctrl+Shift+F9 键删除之前添加的所有断点。
(3) 在 case4 函数第 1 行代码处添加一个断点。
(4) 按 F5 键启动调试项目。此时程序会在刚刚添加的断点处中断。
(5) 选择"调试"菜单"窗口"中的"反汇编",打开"反汇编"窗口。

在打开的"反汇编"窗口中显示了 case4 函数的反汇编代码,如下面的代码清单所示(为了方便对代码进行说明,添加了行号):

```
1)  push  ebp
2)  mov   ebp,esp
3)  sub   esp,0x8
4)  mov   DWORD PTR [esp],0xc
5)  call  0x40195c<malloc>
6)  mov   DWORD PTR [ebp-0x4],eax
7)  mov   edx,DWORD PTR [ebp-0x4]
8)  mov   eax,DWORD PTR [ebp-0x4]
9)  mov   DWORD PTR [edx+0x4],eax
10) mov   eax,DWORD PTR [ebp-0x4]
11) mov   eax,DWORD PTR [eax+0x8]
12) mov   DWORD PTR [ebp-0x4],eax
13) leave
14) ret
```

下面对汇编代码进行详细说明:

(1) 第 3 行代码在栈上开辟了一个 8 字节的空间,用来存放当前函数的临时变量 p 和传给 malloc 函数的参数。

(2) 第 4 行代码将传给 malloc 函数的参数 0xc(TreeNode 结构体的大小为 12 个字节)放入栈顶。

(3) 第 5 行代码调用 malloc 函数。

（4）第 6 行代码将 malloc 函数的返回值从 eax 寄存器中复制到临时变量 p 中，这样变量 p 就指向了由 malloc 函数在堆中刚刚分配的 12 个字节的基址。堆中的 12 个字节是这样分配的：域 val 占用了最低的 4 个字节，域 lchild 占用了中间的 4 个字节，域 rchild 占用了最高的 4 个字节。此时的堆和栈如图 12-5 所示。

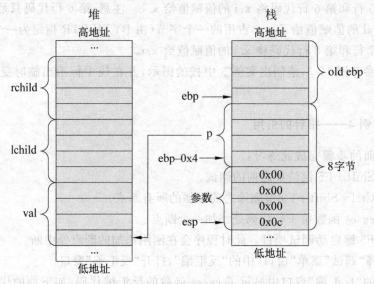

图 12-5　调用 malloc 函数后的堆和栈

（5）第 7 行将变量 p 的值赋值给了 edx，这样 edx 也就指向了堆中 12 字节的基址。第 8 行又将变量 p 的值赋值给了 eax。第 9 行使用 edx 为基址，加 4 后就是 lchild 在堆中的位置，从而完成了将变量 p 的值赋值给 p->lchild，此时 p->lchild 也就指向了堆中 12 字节的基址。

（6）与之前的代码类似，第 10、11、12 行将 p->rchild 的值赋值给变量 p，此时 p 就不再指向之前的位置了，而是指向一个随机的位置（因为 p->rchild 没有初始化）。

3.7　案例 5——if 语句

按照下面的步骤完成此练习：

（1）按 Shift+F5 键结束之前的调试。

（2）按 Ctrl+Shift+F9 键删除之前添加的所有断点。

（3）在 case5 函数第 1 行代码处添加一个断点。

（4）按 F5 键启动调试项目。此时程序会在刚刚添加的断点处中断。

（5）选择"调试"菜单"窗口"中的"反汇编"，打开"反汇编"窗口。

在打开的"反汇编"窗口中显示了 case5 函数的反汇编代码，如下面的代码清单所示（为了方便对代码进行说明，保留了每行指令相对于函数起始地址的偏移）：

```
<case5+0>:      push    ebp
<case5+1>:      mov     ebp,esp
<case5+3>:      sub     esp,0x8
```

```
<case5+6>:     mov    eax,DWORD PTR [ebp-0x4]
<case5+9>:     cmp    eax,DWORD PTR [ebp-0x8]
<case5+12>:    jle    0x4013e5<case5+21>
<case5+14>:    lea    eax,[ebp-0x8]
<case5+17>:    inc    DWORD PTR [eax]
<case5+19>:    jmp    0x4013ea<case5+26>
<case5+21>:    lea    eax,[ebp-0x4]
<case5+24>:    dec    DWORD PTR [eax]
<case5+26>:    leave
<case5+27>:    ret
```

下面对汇编代码进行详细说明：

（1）偏移为 3 的代码在栈上开辟了一个 8 字节的空间，用来存放当前函数的临时变量 x 和 y，其中，变量 x 的位置是 ebp-0x4，变量 y 的位置是 ebp-0x8。

（2）偏移为 6 和 9 的代码比较 x 和 y。

（3）偏移为 12 的代码行判断当 $x \leqslant y$ 时，就跳转执行偏移为 21 和 24 的代码行，完成 x-- 操作后从函数中返回。相当于源代码中的 else 分支。

（4）如果 $x > y$，就会继续执行偏移为 14 和 17 的代码行，完成 y++ 操作。然后再执行偏移为 19 的代码行，跳转到偏移为 26 的代码行后从函数中返回。

四、思考与练习

1. 在案例 1 中研究的是 C 语言默认的函数调用约定（__cdecl），C 语言函数还可以使用其他的调用约定，例如常用的 stdcall、fastcall 等。尝试将 add 函数分别修改为 int __stdcall add(int i, int j) 和 int __fastcall add(int i, int j)，从而显式指定函数的调用约定（注意关键字的前缀是双下画线），然后通过研究不同调用约定下 case1 函数和 add 函数的反汇编代码，总结出这 3 种函数调用约定的异同，并尝试说明哪种调用约定支持可变参数。

2. 编写一个递归函数，在递归调用的过程中，结合其反汇编代码说明在发生递归时栈的构造与展开的过程。

3. 使用下面的函数，结合其反汇编代码说明二维数组中的元素是如何被引用的。

```
void case6()
{
    int i=1;
    int a[3][3];
    a[0][0]=6;
    a[1][1]=7;
    a[1][i+1]=8;
}
```

4. 使用下面的函数，结合其反汇编代码说明联合中的域是如何被引用的。

```
typedef struct _Cub
```

```
    {
        int i;
        union {
            char c;
            int j;
        }u;
    }Cub;
    void case7()
    {
        Cub x;
        x.i=1;
        x.u.c=x.i;
        x.u.j=x.i;
    }
```

5. 使用下面的函数,结合其反汇编代码说明指针的指针是如何被引用的。

```
void case8()
{
    TreeNode * p=(TreeNode * )malloc(sizeof(TreeNode));
    TreeNode * * pp=&p;
    ( * pp)->val=1;
}
```

6. 使用下面的函数,结合其反汇编代码说明 while 语句是如何跳转的。

```
void case9()
{
    int x, y;
    while(x<y)
        y-=x;
}
```

7. 验证 GCC 提供的 C 编译器是否实现了短回路布尔操作。

附录 A TINY 编译器和 TM 机

一、TINY 语言

TINY 语言的程序结构很简单，它在语法上与 Ada 或 PASCAL 相似：仅是一个由分号分隔开的语句序列。另外，它既无过程也无声明。所有的变量都是整型变量，通过对其赋值可较轻易地声明变量。它只有两个控制语句：if 语句和 repeat 语句，这两个控制语句本身也可包含语句序列。if 语句有一个可选的 else 部分且必须由关键字 end 结束。除此之外，read 语句和 write 语句完成输入和输出。在大括号中可以有注释，但注释不能嵌套。

TINY 的表达式也局限于布尔表达式和整型算术表达式。布尔表达式由对两个算术表达式的比较组成，该比较使用 $<$ 与 $=$ 比较算符。算术表达式可以包含整数常量、变量、参数以及 4 个整型算符 $+$、$-$、$*$、$/$，此外还有一般的数学属性。布尔表达式可能只作为测试出现在控制语句中，而没有布尔变量、赋值或 I/O。

下面是该语言中的一个阶乘函数的简单程序示例。

```
{ Sample program
  int TINY language-
  computes factorial
}
read x; { input an integer }
if 0<x then { don't compute if x<=0 }
    fact :=1;
    repeat
        fact :=fact * x;
        x :=x-1
    until
    x=0;
    write fact { output factorial of x }
end
```

二、TINY 编译器

TINY 编译器包含以下的 C 文件：

globals. h(参见源代码清单 tiny-1)　　main. c(参见源代码清单 tiny-9)
util. h(参见源代码清单 tiny-2)　　　util. c(参见源代码清单 tiny-10)

scan. h(参见源代码清单 tiny-3)　　　　scan. c(参见源代码清单 tiny-11)
parse. h(参见源代码清单 tiny-4)　　　parse. c(参见源代码清单 tiny-12)
symtab. h(参见源代码清单 tiny-5)　　symtab. c(参见源代码清单 tiny-13)
analyze. h(参见源代码清单 tiny-6)　　analyze. c(参见源代码清单 tiny-14)
code. h(参见源代码清单 tiny-7)　　　code. c(参见源代码清单 tiny-15)
cgen. h(参见源代码清单 tiny-8)　　　cgen. c(参见源代码清单 tiny-16)

任何代码文件都包含 globals. h 头文件,它包含了数据类型的定义和整个编译器均使用的全局变量。main. c 文件包括运行编译器的主程序,它还分配和初始化全局变量。其他的文件包含了头/代码文件对、在头文件中给出的外部可用的函数原型以及相关代码文件中的实现(包括静态局部变量)。scan、parse、analyze 和 cgen 文件分别对应扫描程序、分析程序、语义分析程序和代码生成器各个阶段。util 文件包括了实用程序函数,生成源代码(语法树)的内部表示和显示列表与出错信息均需要这些函数。symtab 文件包括执行与 TINY 应用相符的符号表的杂凑表。code 文件包括用于依赖目标机器(依赖目标机器在这里指 TM 机,会在稍后做具体介绍)的代码生成的实用程序。

虽然这些文件的交互少了,但是编译器仍有 4 遍:第 1 遍由构造语法树的扫描程序和分析程序组成;第 2 遍和第 3 遍执行语义分析,其中,第 2 遍构造符号表而第 3 遍完成类型检查;最后一遍是代码生成器。

CP Lab 可编译 TINY 编译器,生成的可执行文件是 tiny,通过使用以下命令:

```
tiny sample.txt(参见源代码清单 tiny-17)
```

就可用 TINY 编译器编译文本文件 sample. txt 中的 TINY 源程序,并将目标代码输出到 sample. tm. txt 文件中(在下面将谈到的 TM 机中使用)。

具体的操作可以遵循以下的步骤:

(1) 打开 CP Lab,在"文件"菜单中选择"新建"|"项目",打开"新建项目"对话框。使用模板"TINY 编译器"新建一个项目。

(2) 在"项目管理器"窗口中除 TINY 编译器包含的 C 文件以外,在 TINY 语言样例筛选器下还包含了 sample. txt 和 sample. tm. txt 两个文件,其中,sample. txt 文件是一个从标准输入中读取数据,计算其阶乘后标准输出的 TINY 语言程序,sample. tm. txt 文件用于保存编译后生成的目标代码(TM 机中使用)。

(3) 按 F7 键生成项目。

(4) 按 F5 键启动调试,这时可以看到在 sample. tm . txt 文件中生成了目标代码。

```
                源代码清单 tiny-1: globals.h
/****************************************/
/* File: globals.h                       */
/* Global types and vars for TINY compiler */
/* must come before other include files  */
/* Compiler Construction: Principles and Practice */
/* Kenneth C. Louden                     */
/****************************************/
```

```c
#ifndef _GLOBALS_H_
#define _GLOBALS_H_

#include<stdio.h>
#include<stdlib.h>
#include<ctype.h>
#include<string.h>

#ifndef FALSE
#define FALSE 0
#endif

#ifndef TRUE
#define TRUE 1
#endif

/* MAXRESERVED=the number of reserved words */
#define MAXRESERVED 8

typedef enum
    /* book-keeping tokens */
    {ENDFILE,ERROR,
    /* reserved words */
    IF,THEN,ELSE,END,REPEAT,UNTIL,READ,WRITE,
    /* multicharacter tokens */
    ID,NUM,
    /* special symbols */
    ASSIGN,EQ,LT,PLUS,MINUS,TIMES,OVER,LPAREN,RPAREN,SEMI
    } TokenType;

extern FILE * source; /* source code text file */
extern FILE * listing; /* listing output text file */
extern FILE * code; /* code text file for TM simulator */

extern int lineno; /* source line number for listing */

/**************************************/
/*********** Syntax tree for parsing ****** */
/**************************************/

typedef enum {StmtK,ExpK} NodeKind;
typedef enum {IfK,RepeatK,AssignK,ReadK,WriteK} StmtKind;
typedef enum {OpK,ConstK,IdK} ExpKind;
```

```c
/* ExpType is used for type checking */
typedef enum {Void,Integer,Boolean} ExpType;

#define MAXCHILDREN 3

typedef struct treeNode
    { struct treeNode * child[MAXCHILDREN];
      struct treeNode * sibling;
      int lineno;
      NodeKind nodekind;
      union { StmtKind stmt; ExpKind exp;} kind;
      union { TokenType op;
              int val;
              char * name; } attr;
      ExpType type; /* for type checking of exps */
    } TreeNode;

/***********************************/
/*********** Flags for tracing   ***/
/***********************************/

/* EchoSource=TRUE causes the source program to
 * be echoed to the listing file with line numbers
 * during parsing
 */
extern int EchoSource;

/* TraceScan=TRUE causes token information to be
 * printed to the listing file as each token is
 * recognized by the scanner
 */
extern int TraceScan;

/* TraceParse=TRUE causes the syntax tree to be
 * printed to the listing file in linearized form
 * (using indents for children)
 */
extern int TraceParse;

/* TraceAnalyze=TRUE causes symbol table inserts
 * and lookups to be reported to the listing file
 */
extern int TraceAnalyze;
```

```
/* TraceCode=TRUE causes comments to be written
 * to the TM code file as code is generated
 */
extern int TraceCode;

/* Error=TRUE prevents further passes if an error occurs */
extern int Error;
#endif
```

源代码清单 tiny-2: util.h

```
/****************************************/
/* File: util.h                         */
/* Utility functions for the TINY compiler */
/* Compiler Construction: Principles and Practice */
/* Kenneth C. Louden                    */
/****************************************/

#ifndef _UTIL_H_
#define _UTIL_H_

/* Procedure printToken prints a token
 * and its lexeme to the listing file
 */
void printToken(TokenType, const char * );

/* Function newStmtNode creates a new statement
 * node for syntax tree construction
 */
TreeNode * newStmtNode(StmtKind);

/* Function newExpNode creates a new expression
 * node for syntax tree construction
 */
TreeNode * newExpNode(ExpKind);

/* Function copyString allocates and makes a new
 * copy of an existing string
 */
char * copyString(char * );
```

```
/* procedure printTree prints a syntax tree to the
 * listing file using indentation to indicate subtrees
 */
void printTree(TreeNode * );

#endif
```

源代码清单 tiny-3: scan.h

```
/****************************************/
/* File: scan.h                           */
/* The scanner interface for the TINY compiler */
/* Compiler Construction: Principles and Practice */
/* Kenneth C. Louden                      */
/****************************************/

#ifndef _SCAN_H_
#define _SCAN_H_

/* MAXTOKENLEN is the maximum size of a token */
#define MAXTOKENLEN 40

/* tokenString array stores the lexeme of each token */
extern char tokenString[MAXTOKENLEN+1];

/* function getToken returns the
 * next token in source file
 */
TokenType getToken(void);

#endif
```

源代码清单 tiny-4: parse.h

```
/****************************************/
/* File: parse.h                          */
/* The parser interface for the TINY compiler */
/* Compiler Construction: Principles and Practice */
/* Kenneth C. Louden                      */
/****************************************/

#ifndef _PARSE_H_
#define _PARSE_H_
```

```c
/* Function parse returns the newly
 * constructed syntax tree
 */
TreeNode * parse(void);

#endif
```

源代码清单 tiny-5: symtab.h

```c
/****************************************/
/* File: symtab.h                       */
/* Symbol table interface for the TINY compiler */
/* (allows only one symbol table)       */
/* Compiler Construction: Principles and Practice */
/* Kenneth C. Louden                    */
/****************************************/

#ifndef _SYMTAB_H_
#define _SYMTAB_H_

/* Procedure st_insert inserts line numbers and
 * memory locations into the symbol table
 * loc=memory location is inserted only the
 * first time, otherwise ignored
 */
void st_insert(char * name, int lineno, int loc);

/* Function st_lookup returns the memory
 * location of a variable or-1 if not found
 */
int st_lookup (char * name);

/* Procedure printSymTab prints a formatted
 * listing of the symbol table contents
 * to the listing file
 */
void printSymTab(FILE * listing);

#endif
```

源代码清单 tiny-6: analyze.h

```
/****************************************/
/* File: analyze.h                       */
/* Semantic analyzer interface for TINY compiler */
/* Compiler Construction: Principles and Practice */
/* Kenneth C. Louden                     */
/****************************************/

#ifndef _ANALYZE_H_
#define _ANALYZE_H_

/* Function buildSymtab constructs the symbol
 * table by preorder traversal of the syntax tree
 */
void buildSymtab(TreeNode *);

/* Procedure typeCheck performs type checking
 * by a postorder syntax tree traversal
 */
void typeCheck(TreeNode *);

#endif
```

源代码清单 tiny-7: code.h

```
/****************************************/
/* File: code.h                          */
/* Code emitting utilities for the TINY compiler */
/* and interface to the TM machine       */
/* Compiler Construction: Principles and Practice */
/* Kenneth C. Louden                     */
/****************************************/

#ifndef _CODE_H_
#define _CODE_H_

/* pc=program counter */
#define pc 7

/* mp="memory pointer" points
 * to top of memory (for temp storage)
 */
#define mp 6
```

```
/* gp="global pointer" points
 * to bottom of memory for (global)
 * variable storage
 */
#define gp 5

/* accumulator */
#define ac 0

/* 2nd accumulator */
#define ac1 1

/* code emitting utilities */

/* Procedure emitComment prints a comment line
 * with comment c in the code file
 */
void emitComment(char * c);

/* Procedure emitRO emits a register-only
 * TM instruction
 * op=the opcode
 * r=target register
 * s=1st source register
 * t=2nd source register
 * c=a comment to be printed if TraceCode is TRUE
 */
void emitRO(char * op, int r, int s, int t, char * c);

/* Procedure emitRM emits a register-to-memory
 * TM instruction
 * op=the opcode
 * r=target register
 * d=the offset
 * s=the base register
 * c=a comment to be printed if TraceCode is TRUE
 */
void emitRM(char * op, int r, int d, int s, char * c);

/* Function emitSkip skips "howMany" code
 * locations for later backpatch. It also
 * returns the current code position
 */
```

```
int emitSkip(int howMany);

/* Procedure emitBackup backs up to
 * loc=a previously skipped location
 */
void emitBackup(int loc);

/* Procedure emitRestore restores the current
 * code position to the highest previously
 * unemitted position
 */
void emitRestore(void);

/* Procedure emitRM_Abs converts an absolute reference
 * to a pc-relative reference when emitting a
 * register-to-memory TM instruction
 * op=the opcode
 * r=target register
 * a=the absolute location in memory
 * c=a comment to be printed if TraceCode is TRUE
 */
void emitRM_Abs(char * op, int r, int a, char * c);

#endif
```

源代码清单 tiny-8: cgen.h

```
/****************************************/
/* File: cgen.h                         */
/* The code generator interface to the TINY compiler*/
/* Compiler Construction: Principles and Practice  */
/* Kenneth C. Louden                    */
/****************************************/

#ifndef _CGEN_H_
#define _CGEN_H_

/* Procedure codeGen generates code to a code
 * file by traversal of the syntax tree. The
 * second parameter (codefile) is the file name
 * of the code file, and is used to print the
```

```
 * file name as a comment in the code file
 */
void codeGen(TreeNode * syntaxTree, char * codefile);

#endif
```

源代码清单 tiny-9: main.c

```
/****************************************/
/* File: main.c                         */
/* Main program for TINY compiler       */
/* Compiler Construction: Principles and Practice */
/* Kenneth C. Louden                    */
/****************************************/

#include "globals.h"

/* set NO_PARSE to TRUE to get a scanner-only compiler */
#define NO_PARSE FALSE
/* set NO_ANALYZE to TRUE to get a parser-only compiler */
#define NO_ANALYZE FALSE

/* set NO_CODE to TRUE to get a compiler that does not
 * generate code
 */
#define NO_CODE FALSE

#include "util.h"
#if NO_PARSE
#include "scan.h"
#else
#include "parse.h"
#if !NO_ANALYZE
#include "analyze.h"
#if !NO_CODE
#include "cgen.h"
#endif
#endif
#endif

/* allocate global variables */ int lineno=0;
FILE * source;
```

```c
FILE * listing;
FILE * code;

/* allocate and set tracing flags */
int EchoSource=FALSE;
int TraceScan=FALSE;
int TraceParse=FALSE;
int TraceAnalyze=FALSE;
int TraceCode=FALSE;

int Error=FALSE;
char pgm[256]; /* source code file name */

main(int argc, char * argv[])
{
    TreeNode * syntaxTree;

    if (argc!=2)
    {
        fprintf(stderr,"usage: %s<filename>\n",argv[0]);
        exit(1);
    }
    strcpy(pgm, argv[1]);
    if (strchr (pgm, '.')==NULL)
       strcat(pgm,".tny");
    source=fopen(pgm,"r");
    if (source==NULL)
    { fprintf(stderr,"File %s not found\n",pgm);
      exit(1);
    }
    listing=stdout; /* send listing to screen */
    fprintf(listing,"\nTINY COMPILATION: %s\n",pgm);
#if NO_PARSE
    while (getToken()!=ENDFILE);
#else
    syntaxTree=parse();
    if (TraceParse) {
      fprintf(listing,"\nSyntax tree:\n");
      printTree(syntaxTree);
    }
#if!NO_ANALYZE
    if (! Error)
```

```
    { if (TraceAnalyze) fprintf(listing,"\nBuilding Symbol Table...\n");
      buildSymtab(syntaxTree);
      if (TraceAnalyze) fprintf(listing,"\nChecking Types...\n");
      typeCheck(syntaxTree);
      if (TraceAnalyze) fprintf(listing,"\nType Checking Finished\n");
    }
#if!NO_CODE
    if (! Error)
    { char * codefile;
      int fnlen=strcspn(pgm,".");
      codefile=(char *) calloc(fnlen+4, sizeof(char));
      strncpy(codefile,pgm,fnlen);
      strcat(codefile,".tm.txt");
      code=fopen(codefile,"w");
      if (code==NULL)
      { printf("Unable to open %s\n",codefile);
        exit(1);
      }
      codeGen(syntaxTree,codefile);
      fclose(code);
    }
#endif
#endif
#endif
    fclose(source);
    return 0;
}
```

源代码清单 tiny-10: util.c

```
/* * * * * * * * * * * * * * * * * * * * * * * * * */
/* File: util.c                                    */
/* Utility function implementation                 */
/* for the TINY compiler                           */
/* Compiler Construction: Principles and Practice  */
/* Kenneth C. Louden                               */
/* * * * * * * * * * * * * * * * * * * * * * * * * */

#include "globals.h"
#include "util.h"

/* Procedure printToken prints a token
 * and its lexeme to the listing file
 */
```

```
void printToken(TokenType token, const char * tokenString)
{
    switch (token)
    { case IF:
      case THEN:
      case ELSE:
      case END:
      case REPEAT:
      case UNTIL:
      case READ:
      case WRITE:
        fprintf(listing,
            "reserved word: %s\n",tokenString);
        break;
      case ASSIGN: fprintf(listing,":=\n"); break;
      case LT: fprintf(listing,"<\n"); break;
      case EQ: fprintf(listing,"=\n"); break;
      case LPAREN: fprintf(listing,"(\n"); break;
      case RPAREN: fprintf(listing,")\n"); break;
      case SEMI: fprintf(listing,";\n"); break;
      case PLUS: fprintf(listing,"+\n"); break;
      case MINUS: fprintf(listing,"-\n"); break;
      case TIMES: fprintf(listing,"*\n"); break;
      case OVER: fprintf(listing,"/\n"); break;
      case ENDFILE: fprintf(listing,"EOF\n"); break;
      case NUM:
        fprintf(listing,
            "NUM, val=%s\n",tokenString);
        break;
      case ID:
        fprintf(listing,
            "ID, name=%s\n",tokenString);
        break;
      case ERROR:
        fprintf(listing,
            "ERROR: %s\n",tokenString);
        break;
      default: /* should never happen */
        fprintf(listing,"Unknown token: %d\n",token);
    }
}
```

```
/* Function newStmtNode creates a new statement
 * node for syntax tree construction
 */
TreeNode * newStmtNode(StmtKind kind)
{ TreeNode * t=(TreeNode *) malloc(sizeof(TreeNode));
  int i;
  if (t==NULL)
    fprintf(listing,"Out of memory error at line %d\n",lineno);
  else {
    for (i=0;i<MAXCHILDREN;i++) t->child[i]=NULL;
    t->sibling=NULL;
    t->nodekind=StmtK;
    t->kind.stmt=kind;
    t->lineno=lineno;
  }
  return t;
}

/* Function newExpNode creates a new expression
 * node for syntax tree construction
 */
TreeNode * newExpNode(ExpKind kind)
{ TreeNode * t=(TreeNode *) malloc(sizeof(TreeNode));
  int i;
  if (t==NULL)
    fprintf(listing,"Out of memory error at line %d\n",lineno);
  else {
    for (i=0;i<MAXCHILDREN;i++) t->child[i]=NULL;
    t->sibling=NULL;
    t->nodekind=ExpK;
    t->kind.exp=kind;
    t->lineno=lineno;
    t->type=Void;
  }
  return t;
}

/* Function copyString allocates and makes a new
 * copy of an existing string
 */
char * copyString(char * s)
{ int n;
  char * t;
```

```
    if (s==NULL) return NULL;
  n=strlen(s)+1;
  t=malloc(n);
  if (t==NULL)
    fprintf(listing,"Out of memory error at line %d\n",lineno);
  else strcpy(t,s);
  return t;
}

/* Variable indentno is used by printTree to
 * store current number of spaces to indent
 */
static indentno=0;

/* macros to increase/decrease indentation */
#define INDENT indentno+=2
#define UNINDENT indentno-=2

/* printSpaces indents by printing spaces */
static void printSpaces(void)
{ int i;
  for (i=0;i<indentno;i++)
    fprintf(listing," ");
}

/* procedure printTree prints a syntax tree to the
 * listing file using indentation to indicate subtrees
 */
void printTree(TreeNode * tree)
{ int i;
  INDENT;
  while (tree!=NULL) {
     printSpaces();
     if (tree->nodekind==StmtK)
     { switch (tree->kind.stmt) {
         case IfK:
           fprintf(listing,"If\n");
           break;
         case RepeatK:
           fprintf(listing,"Repeat\n");
           break;
         case AssignK:
           fprintf(listing,"Assign to: %s\n",tree->attr.name);
```

```
          break;
        case ReadK:
          fprintf(listing,"Read: %s\n",tree->attr.name);
          break;
        case WriteK:
          fprintf(listing,"Write\n");
          break;
        default:
          fprintf(listing,"Unknown ExpNode kind\n");
          break;
       }
    }
    else if (tree->nodekind==ExpK)
    { switch (tree->kind.exp) {
        case OpK:
          fprintf(listing,"Op: ");
          printToken(tree->attr.op,"\0");
          break;
        case ConstK:
          fprintf(listing,"Const: %d\n",tree->attr.val);
          break;
        case IdK:
          fprintf(listing,"Id: %s\n",tree->attr.name);
          break;
        default:
          fprintf(listing,"Unknown ExpNode kind\n");
          break;
      }
    }
    else fprintf(listing,"Unknown node kind\n");
    for (i=0;i<MAXCHILDREN;i++)
         printTree(tree->child[i]);
    tree=tree->sibling;
  }
  UNINDENT;
}
```

源代码清单 tiny-11: scan.c

```
/****************************************/
/* File: scan.c                         */
/* The scanner implementation for the TINY compiler */
/* Compiler Construction: Principles and Practice   */
/* Kenneth C. Louden                    */
/****************************************/
```

```c
#include "globals.h"
#include "util.h"
#include "scan.h"

/* states in scanner DFA */
typedef enum
    { START,INASSIGN,INCOMMENT,INNUM,INID,DONE }
    StateType;

/* lexeme of identifier or reserved word */
char tokenString[MAXTOKENLEN+1];

/* BUFLEN=length of the input buffer for
source code lines */
#define BUFLEN 256

static char lineBuf[BUFLEN];        /* holds the current line */
static int linepos=0;               /* current position in LineBuf */
static int bufsize=0;               /* current size of buffer string */
static int EOF_flag=FALSE;          /* corrects ungetNextChar behavior on EOF */

/* getNextChar fetches the next non-blank character
   from lineBuf, reading in a new line if lineBuf is
   exhausted */
static int getNextChar(void)
{ if (!(linepos<bufsize))
  { lineno++;
     if (fgets(lineBuf,BUFLEN-1,source))
     { if (EchoSource) fprintf(listing,"%4d: %s",lineno,lineBuf);
       bufsize=strlen(lineBuf);
       linepos=0;
       return lineBuf[linepos++];
     }
     else
     { EOF_flag=TRUE;
       return EOF;
     }
  }
  else return lineBuf[linepos++];
}

/* ungetNextChar backtracks one character
   in lineBuf */
```

```
static void ungetNextChar(void)
{ if (!EOF_flag) linepos--;}

/* lookup table of reserved words */
static struct
    { char * str;
      TokenType tok;
    } reservedWords[MAXRESERVED]
={{"if",IF},{"then",THEN},{"else",ELSE},{"end",END},
  {"repeat",REPEAT},{"until",UNTIL},{"read",READ},
  {"write",WRITE}};

/* lookup an identifier to see if it is a reserved word */
/* uses linear search */
static TokenType reservedLookup (char * s)
{ int i;
  for (i=0;i<MAXRESERVED;i++)
     if (!strcmp(s,reservedWords[i].str))
        return reservedWords[i].tok;
  return ID;
}

/****************************************/
/* the primary function of the scanner */
/****************************************/
/* function getToken returns the
 * next token in source file
 */
TokenType getToken(void)
{  /* index for storing into tokenString */
   int tokenStringIndex=0;
   /* holds current token to be returned */
   TokenType currentToken;
   /* current state-always begins at START */
   StateType state=START;
   /* flag to indicate save to tokenString */
   int save;
   while (state!=DONE)
   { int c=getNextChar();
     save=TRUE;
     switch (state)
     { case START:
          if (isdigit(c))
```

```
            state=INNUM;
        else if (isalpha(c))
            state=INID;
        else if (c==':')
            state=INASSIGN;
        else if ((c==' ') || (c=='\t') || (c=='\n'))
            save=FALSE;
        else if (c=='{')
        { save=FALSE;
          state=INCOMMENT;
        }
        else
        { state=DONE;
          switch (c)
          { case EOF:
              save=FALSE;
              currentToken=ENDFILE;
              break;
            case '=':
              currentToken=EQ;
              break;
            case '<':
              currentToken=LT;
              break;
            case '+':
              currentToken=PLUS;
              break;
            case '-':
              currentToken=MINUS;
              break;
            case '*':
              currentToken=TIMES;
              break;
            case '/':
              currentToken=OVER;
              break;
            case '(':
              currentToken=LPAREN;
              break;
            case ')':
              currentToken=RPAREN;
              break;
            case ';':
```

```
            currentToken=SEMI;
            break;
          default:
            currentToken=ERROR;
            break;
        }
    }
      break;
    case INCOMMENT:
      save=FALSE;
      if (c==EOF)
      { state=DONE;
        currentToken=ENDFILE;
      }
      else if (c=='}') state=START;
      break;
    case INASSIGN:
      state=DONE;
      if (c=='=')
        currentToken=ASSIGN;
      else
      { /* backup in the input */
        ungetNextChar();
        save=FALSE;
        currentToken=ERROR;
      }
      break;
    case INNUM:
      if (!isdigit(c))
      { /* backup in the input */
        ungetNextChar();
        save=FALSE;
        state=DONE;
        currentToken=NUM;
      }
      break;
    case INID:
      if (!isalpha(c))
      { /* backup in the input */
        ungetNextChar();
        save=FALSE;
        state=DONE;
        currentToken=ID;
      }
      break;
```

```
        case DONE:
        default: /* should never happen */
          fprintf(listing,"Scanner Bug: state=%d\n",state);
          state=DONE;
          currentToken=ERROR;
          break;
      }
      if ((save) && (tokenStringIndex<=MAXTOKENLEN))
        tokenString[tokenStringIndex++]=(char) c;
      if (state==DONE)
        { tokenString[tokenStringIndex]='\0';
          if (currentToken==ID)
            currentToken=reservedLookup(tokenString);
        }
    }
    if (TraceScan) {
      fprintf(listing,"\t%d: ",lineno);
      printToken(currentToken,tokenString);
    }
    return currentToken;
} /* end getToken */
```

源代码清单 tiny-12: **parse.c**

```
/****************************************/
/* File: parse.c                         */
/* The parser implementation for the TINY compiler */
/* Compiler Construction: Principles and Practice */
/* Kenneth C. Louden                     */
/****************************************/

#include "globals.h"
#include "util.h"
#include "scan.h"
#include "parse.h"

static TokenType token; /* holds current token */

/* function prototypes for recursive calls */
static TreeNode * stmt_sequence(void);
static TreeNode * statement(void);
static TreeNode * if_stmt(void);
static TreeNode * repeat_stmt(void);
```

```c
static TreeNode * assign_stmt(void);
static TreeNode * read_stmt(void);
static TreeNode * write_stmt(void);
static TreeNode * exp(void);
static TreeNode * simple_exp(void);
static TreeNode * term(void);
static TreeNode * factor(void);
extern char pgm[256];

//G:\Documents and Settings\xjx\My Documents\CP Lab\Projects\trter\main.c:
40: error: syntax error before "int"
//Syntax error at line 11: unexpected token->reserved word: until
static void syntaxError(char * message)
{
    fprintf(listing,"\n%s:%d: error: %s",pgm, lineno, message);
    Error=TRUE;
}

static void match(TokenType expected)
{ if (token==expected) token=getToken();
  else {
      syntaxError("unexpected token->");
      printToken(token,tokenString);
      fprintf(listing,"      ");
  }
}

TreeNode * stmt_sequence(void)
{ TreeNode * t=statement();
  TreeNode * p=t;
  while ((token!=ENDFILE) && (token!=END) &&
         (token!=ELSE) && (token!=UNTIL))
  { TreeNode * q;
    match(SEMI);
    q=statement();
    if (q!=NULL) {
      if (t==NULL) t=p=q;
      else /* now p cannot be NULL either */
      { p->sibling=q;
        p=q;
      }
    }
  }
  return t;
}
```

```
TreeNode * statement(void)
{ TreeNode * t=NULL;
  switch (token) {
    case IF : t=if_stmt(); break;
    case REPEAT : t=repeat_stmt(); break;
    case ID : t=assign_stmt(); break;
    case READ : t=read_stmt(); break;
    case WRITE : t=write_stmt(); break;
    default: syntaxError("unexpected token->");
            printToken(token,tokenString);
            token=getToken();
            break;
  } /* end case */
  return t;
}

TreeNode * if_stmt(void)
{ TreeNode * t=newStmtNode(IfK);
  match(IF);
  if (t!=NULL) t->child[0]=exp();
  match(THEN);
  if (t!=NULL) t->child[1]=stmt_sequence();
  if (token==ELSE) {
     match(ELSE);
     if (t!=NULL) t->child[2]=stmt_sequence();
  }
  match(END);
  return t;
}

TreeNode * repeat_stmt(void)
{ TreeNode * t=newStmtNode(RepeatK);
  match(REPEAT);
  if (t!=NULL) t->child[0]=stmt_sequence();
  match(UNTIL);
  if (t!=NULL) t->child[1]=exp();
  return t;
}

TreeNode * assign_stmt(void)
{ TreeNode * t=newStmtNode(AssignK);
  if ((t!=NULL) && (token==ID))
    t->attr.name=copyString(tokenString);
```

```
  match(ID);
  match(ASSIGN);
  if (t!=NULL) t->child[0]=exp();
  return t;
}

TreeNode * read_stmt(void)
{ TreeNode * t=newStmtNode(ReadK);
  match(READ);
  if ((t!=NULL) && (token==ID))
    t->attr.name=copyString(tokenString);
  match(ID);
  return t;
}

TreeNode * write_stmt(void)
{ TreeNode * t=newStmtNode(WriteK);
  match(WRITE);
  if (t!=NULL) t->child[0]=exp();
  return t;
}

static TreeNode * exp(void)
{ TreeNode * t=simple_exp();
  if ((token==LT)||(token==EQ)) {
    TreeNode * p=newExpNode(OpK);
    if (p!=NULL) {
      p->child[0]=t;
      p->attr.op=token;
      t=p;
      }
    match(token);
    if (t!=NULL)
       t->child[1]=simple_exp();
  }
  return t;
}

TreeNode * simple_exp(void)
{ TreeNode * t=term();
  while ((token==PLUS)||(token==MINUS))
  { TreeNode * p=newExpNode(OpK);
    if (p!=NULL) {
      p->child[0]=t;
```

```
          p->attr.op=token;
          t=p;
          match(token);
          t->child[1]=term();
        }
    }
    return t;
}

TreeNode * term(void)
{ TreeNode * t=factor();
  while ((token==TIMES)||(token==OVER))
    { TreeNode * p=newExpNode(OpK);
      if (p!=NULL) {
         p->child[0]=t;
         p->attr.op=token;
         t=p;
         match(token);
         p->child[1]=factor();
      }
    }
    return t;
}

TreeNode * factor(void)
{ TreeNode * t=NULL;
  switch (token) {
    case NUM :
      t=newExpNode(ConstK);
      if ((t!=NULL) && (token==NUM))
        t->attr.val=atoi(tokenString);
      match(NUM);
      break;
    case ID:
      t=newExpNode(IdK);
      if ((t!=NULL) && (token==ID))
         t->attr.name=copyString(tokenString);
      match(ID);
      break;
    case LPAREN :
      match(LPAREN);
      t=exp();
      match(RPAREN);
```

```
           break;
        default:
          syntaxError("unexpected token->");
          printToken(token,tokenString);
          token=getToken();
          break;
      }
   return t;
}

/****************************/
/* the primary function of the parser */
/****************************/
/* Function parse returns the newly
 * constructed syntax tree
 */
TreeNode * parse(void)
{  TreeNode * t;
   token=getToken();
   t=stmt_sequence();
   if (token!=ENDFILE)
      syntaxError("Code ends before file\n");
   return t;
}
```

源代码清单 tiny-13: symtab.c

```
/****************************/
/* File: symtab.c                            */
/* Symbol table implementation for the TINY compiler */
/* (allows only one symbol table)            */
/* Symbol table is implemented as a chained  */
/* hash table                                */
/* Compiler Construction: Principles and Practice */
/* Kenneth C. Louden                         */
/****************************/

#include<stdio.h>
#include<stdlib.h>
#include<string.h>
#include "symtab.h"

/* SIZE is the size of the hash table */
```

```
#define SIZE 211

/* SHIFT is the power of two used as multiplier
   in hash function */
#define SHIFT 4

/* the hash function */
static int hash (char * key)
{  int temp=0;
   int i=0;
   while (key[i]!='\0')
   {  temp=((temp<<SHIFT)+key[i]) %SIZE;
      ++i;
   }
   return temp;
}

/* the list of line numbers of the source
 * code in which a variable is referenced
 */
typedef struct LineListRec
   {  int lineno;
      struct LineListRec * next;
   } * LineList;
```

```
/* The record in the bucket lists for
 * each variable, including name,
 * assigned memory location, and
 * the list of line numbers in which
 * it appears in the source code
 */
typedef struct BucketListRec
   { char * name;
     LineList lines;
     int memloc; /* memory location for variable */
     struct BucketListRec * next;
   } * BucketList;

/* the hash table */
static BucketList hashTable[SIZE];
```

```c
/* Procedure st_insert inserts line numbers and
 * memory locations into the symbol table
 * loc=memory location is inserted only the
 * first time, otherwise ignored
 */
void st_insert(char * name, int lineno, int loc)
{  int h=hash(name);
   BucketList l=hashTable[h];
   while ((l!=NULL) && (strcmp(name,l->name)!=0))
     l=l->next;
   if (l==NULL) /* variable not yet in table */
   { l=(BucketList) malloc(sizeof(struct BucketListRec));
     l->name=name;
     l->lines= (LineList) malloc(sizeof(struct LineListRec));
     l->lines->lineno=lineno;
     l->memloc=loc;
     l->lines->next=NULL;
     l->next=hashTable[h];
     hashTable[h]=l; }
   else /* found in table, so just add line number */
   {  LineList t=l->lines;
      while (t->next!=NULL) t=t->next;
      t->next= (LineList) malloc(sizeof(struct LineListRec));
      t->next->lineno=lineno;
      t->next->next=NULL;
   }
} /* st_insert */

/* Function st_lookup returns the memory
 * location of a variable or-1 if not found
 */
int st_lookup (char * name)
{  int h=hash(name);
   BucketList l=hashTable[h];
   while ((l!=NULL) && (strcmp(name,l->name)!=0))
     l=l->next;
   if (l==NULL) return-1;
   else return l->memloc;
}

/* Procedure printSymTab prints a formatted
 * listing of the symbol table contents
 * to the listing file
 */
```

```
void printSymTab(FILE * listing)
{ int i;
  fprintf(listing,"Variable Name  Location   Line Numbers\n");
  fprintf(listing,"-------------------------------------\n");
  for (i=0;i<SIZE;++i)
  { if (hashTable[i]!=NULL)
       { BucketList l=hashTable[i];
         while (l!=NULL)
         { LineList t=l->lines;
           fprintf(listing,"%-14s ",l->name);
           fprintf(listing,"%-8d  ",l->memloc);
           while (t!=NULL)
           { fprintf(listing,"%4d ",t->lineno);
             t=t->next;
           }
           fprintf(listing,"\n");
           l=l->next;
         }
       }
  }
} /* printSymTab */
```

源代码清单 tiny-14: analyze.c

```
/****************************************/
/* File: analyze.c                      */
/* Semantic analyzer implementation     */
/* for the TINY compiler                */
/* Compiler Construction: Principles and Practice */
/* Kenneth C. Louden                    */
/****************************************/

#include "globals.h"
#include "symtab.h"
#include "analyze.h"

/* counter for variable memory locations */
static int location=0;

/* Procedure traverse is a generic recursive
 * syntax tree traversal routine:
 * it applies preProc in preorder and postProc
 * in postorder to tree pointed to by t
 */
```

```
static void traverse(TreeNode * t,
             void (* preProc) (TreeNode *),
             void (* postProc) (TreeNode *))
{ if (t!=NULL)
   { preProc(t);
        { int i;
          for (i=0; i<MAXCHILDREN; i++)
            traverse(t->child[i],preProc,postProc);
     }
     postProc(t);
     traverse(t->sibling,preProc,postProc);
   }
}

/* nullProc is a do-nothing procedure to
 * generate preorder-only or postorder-only
 * traversals from traverse
 */
static void nullProc(TreeNode * t)
{ if (t==NULL) return;
  else return;
}

/* Procedure insertNode inserts
 * identifiers stored in t into
 * the symbol table
 */
static void insertNode(TreeNode * t)
{ switch (t->nodekind)
   { case StmtK:
       switch (t->kind.stmt)
        { case AssignK:
          case ReadK:
            if (st_lookup(t->attr.name)==-1)
            /* not yet in table, so treat as new definition */
              st_insert(t->attr.name,t->lineno,location++);
            else
            /* already in table, so ignore location,
               add line number of use only */
              st_insert(t->attr.name,t->lineno,0);
            break;
          default:
            break;
        }
```

```
              break;
        case ExpK:
          switch (t->kind.exp)
          {  case IdK:
               if (st_lookup(t->attr.name)==-1)
               /* not yet in table, so treat as new definition */
                 st_insert(t->attr.name,t->lineno,location++);
               else
               /* already in table, so ignore location,
                          add line number of use only */
                        st_insert(t->attr.name,t->lineno,0);
               break;
            default:
               break;
          }
          break;
        default:
          break;
      }
}

/* Function buildSymtab constructs the symbol
 * table by preorder traversal of the syntax tree
 */
void buildSymtab(TreeNode * syntaxTree)
{ traverse(syntaxTree,insertNode,nullProc);
  if (TraceAnalyze)
  { fprintf(listing,"\nSymbol table:\n\n");
    printSymTab(listing);
  }
}

static void typeError(TreeNode * t, char * message)
{ fprintf(listing,"Type error at line %d: %s\n",t->lineno,message);
  Error=TRUE;
}

/* Procedure checkNode performs
 * type checking at a single tree node
 */
static void checkNode(TreeNode * t)
```

```c
{ switch (t->nodekind)
  { case ExpK:
      switch (t->kind.exp)
      { case OpK:
          if ((t->child[0]->type!=Integer) ||
              (t->child[1]->type!=Integer))
            typeError(t,"Op applied to non-integer");
          if ((t->attr.op==EQ) || (t->attr.op==LT))
            t->type=Boolean;
          else
            t->type=Integer;
          break;
        case ConstK:
        case IdK:
          t->type=Integer;
          break;
        default:
          break;
      }
      break;
    case StmtK:
      switch (t->kind.stmt)
      { case IfK:
          if (t->child[0]->type==Integer)
            typeError(t->child[0],"if test is not Boolean");
          break;
        case AssignK:
          if (t->child[0]->type!=Integer)
            typeError(t->child[0],"assignment of non-integer value");
          break;
        case WriteK:
          if (t->child[0]->type!=Integer)
            typeError(t->child[0],"write of non-integer value");
          break;
        case RepeatK:
          if (t->child[1]->type==Integer)
            typeError(t->child[1],"repeat test is not Boolean");
          break;
        default:
          break;
      }
      break;
    default:
```

```
        break;
    }
}

/* Procedure typeCheck performs type checking
 * by a postorder syntax tree traversal
 */
void typeCheck(TreeNode * syntaxTree)
{
    traverse(syntaxTree,nullProc,checkNode);
}
```

源代码清单 tiny-15: code.c

```
/****************************/
/* File: code.c                               */
/* TM Code emitting utilities                 */
/* implementation for the TINY compiler       */
/* Compiler Construction: Principles and Practice */
/* Kenneth C. Louden                          */
/****************************/

#include "globals.h"
#include "code.h"

/* TM location number for current instruction emission */
static int emitLoc=0;

/* Highest TM location emitted so far
   For use in conjunction with emitSkip,
   emitBackup, and emitRestore */
static int highEmitLoc=0;

/* Procedure emitComment prints a comment line
 * with comment c in the code file
 */
void emitComment(char * c)
{ if (TraceCode) fprintf(code,"* %s\n",c);}

/* Procedure emitRO emits a register-only
 * TM instruction
 * op=the opcode
 * r=target register
```

```c
 * s=1st source register
 * t=2nd source register
 * c=a comment to be printed if TraceCode is TRUE
 */
void emitRO(char * op, int r, int s, int t, char * c)
{   fprintf(code,"%3d:  %5s  %d,%d,%d ",emitLoc++,op,r,s,t);
    if (TraceCode) fprintf(code,"\t%s",c);
    fprintf(code,"\n");
    if (highEmitLoc<emitLoc) highEmitLoc=emitLoc;
} /* emitRO */

/* Procedure emitRM emits a register-to-memory
 * TM instruction
 * op=the opcode
 * r=target register
 * d=the offset
 * s=the base register
 * c=a comment to be printed if TraceCode is TRUE
 */
void emitRM(char * op, int r, int d, int s, char * c)
{   fprintf(code,"%3d:  %5s  %d,%d(%d) ",emitLoc++,op,r,d,s);
    if (TraceCode) fprintf(code,"\t%s",c);
    fprintf(code,"\n");
    if (highEmitLoc<emitLoc) highEmitLoc=emitLoc;
} /* emitRM */

/* Function emitSkip skips "howMany" code
 * locations for later backpatch. It also
 * returns the current code position
 */
int emitSkip(int howMany)
{   int i=emitLoc;
    emitLoc +=howMany;
    if (highEmitLoc<emitLoc) highEmitLoc=emitLoc;
    return i;
} /* emitSkip */

/* Procedure emitBackup backs up to
 * loc=a previously skipped location
 */
void emitBackup(int loc)
{ if (loc>highEmitLoc) emitComment("BUG in emitBackup");
```

```
    emitLoc=loc;
} /* emitBackup */

/* Procedure emitRestore restores the current
 * code position to the highest previously
 * unemitted position
 */
void emitRestore(void)
{   emitLoc=highEmitLoc;}

/* Procedure emitRM_Abs converts an absolute reference
 * to a pc-relative reference when emitting a
 * register-to-memory TM instruction
 * op=the opcode
 * r=target register
 * a=the absolute location in memory
 * c=a comment to be printed if TraceCode is TRUE
 */
void emitRM_Abs(char * op, int r, int a, char * c)
{
    fprintf(code,"%3d:  %5s  %d,%d(%d) ",
            emitLoc,op,r,a-(emitLoc+1),pc);
    ++emitLoc;
    if (TraceCode) fprintf(code,"\t%s",c);
    fprintf(code,"\n");
    if (highEmitLoc<emitLoc) highEmitLoc=emitLoc;
} /* emitRM_Abs */
```

源代码清单 tiny-16: cgen.c

```
/********************************/
/* File: cgen.c                               */
/* The code generator implementation          */
/* for the TINY compiler                      */
/* (generates code for the TM machine)        */
/* Compiler Construction: Principles and Practice */
/* Kenneth C. Louden                          */
/********************************/

#include "globals.h"
#include "symtab.h"
#include "code.h"
#include "cgen.h"
```

```
/* tmpOffset is the memory offset for temps
   It is decremented each time a temp is
   stored, and incremented when loaded again
*/
static int tmpOffset=0;

/* prototype for internal recursive code generator */
static void cGen (TreeNode * tree);

/* Procedure genStmt generates code at a statement node */
static void genStmt(TreeNode * tree)
{ TreeNode *p1, *p2, *p3;
  int savedLoc1,savedLoc2,currentLoc;
  int loc;
  switch (tree->kind.stmt) {

    case IfK :
      if (TraceCode) emitComment("-> if");
      p1=tree->child[0];
      p2=tree->child[1];
      p3=tree->child[2];
      /* generate code for test expression */
      cGen(p1);
      savedLoc1=emitSkip(1);
      emitComment("if: jump to else belongs here");
      /* recurse on then part */
      cGen(p2);
      savedLoc2=emitSkip(1);
      emitComment("if: jump to end belongs here");
      currentLoc=emitSkip(0);
      emitBackup(savedLoc1);
      emitRM_Abs("JEQ",ac,currentLoc,"if: jmp to else");
      emitRestore();
      /* recurse on else part */
      cGen(p3);
      currentLoc=emitSkip(0);
      emitBackup(savedLoc2);
      emitRM_Abs("LDA",pc,currentLoc,"jmp to end");
      emitRestore();
      if (TraceCode) emitComment("<- if");
      break; /* if_k */
```

```
      case RepeatK:
         if (TraceCode) emitComment("-> repeat");
         p1=tree->child[0];
         p2=tree->child[1];
         savedLoc1=emitSkip(0);
         emitComment("repeat: jump after body comes back here");
         /* generate code for body */
         cGen(p1);
         /* generate code for test */
         cGen(p2);
         emitRM_Abs("JEQ",ac,savedLoc1,"repeat: jmp back to body");
         if (TraceCode) emitComment("<- repeat");
         break; /* repeat */

      case AssignK:
         if (TraceCode) emitComment("-> assign");
         /* generate code for rhs */
         cGen(tree->child[0]);
         /* now store value */
         loc=st_lookup(tree->attr.name);
         emitRM("ST",ac,loc,gp,"assign: store value");
         if (TraceCode) emitComment("<- assign");
         break; /* assign_k */

      case ReadK:
         emitRO("IN",ac,0,0,"read integer value");
         loc=st_lookup(tree->attr.name);
         emitRM("ST",ac,loc,gp,"read: store value");
         break;
      case WriteK:
         /* generate code for expression to write */
         cGen(tree->child[0]);
         /* now output it */
         emitRO("OUT",ac,0,0,"write ac");
         break;
      default:
         break;
   }
} /* genStmt */

/* Procedure genExp generates code at an expression node */
static void genExp(TreeNode * tree)
```

```
{ int loc;
  TreeNode *p1, *p2;
  switch (tree->kind.exp) {

    case ConstK :
      if (TraceCode) emitComment("->Const");
      /* gen code to load integer constant using LDC */
      emitRM("LDC",ac,tree->attr.val,0,"load const");
      if (TraceCode) emitComment("<-Const");
      break; /* ConstK */

    case IdK :
      if (TraceCode) emitComment("->Id");
      loc=st_lookup(tree->attr.name);
      emitRM("LD",ac,loc,gp,"load id value");
      if (TraceCode) emitComment("<-Id");
      break; /* IdK */

    case OpK :
      if (TraceCode) emitComment("->Op");
      p1=tree->child[0];
      p2=tree->child[1];
      /* gen code for ac=left arg */
      cGen(p1);
      /* gen code to push left operand */
      emitRM("ST",ac,tmpOffset--,mp,"op: push left");
      /* gen code for ac=right operand */
      cGen(p2);
      /* now load left operand */
      emitRM("LD",ac1,++tmpOffset,mp,"op: load left");
      switch (tree->attr.op) {
          case PLUS :
              emitRO("ADD",ac,ac1,ac,"op +");
              break;
          case MINUS :
              emitRO("SUB",ac,ac1,ac,"op-");
              break;
          case TIMES :
              emitRO("MUL",ac,ac1,ac,"op * ");
              break;
          case OVER :
              emitRO("DIV",ac,ac1,ac,"op /");
              break;
```

```
              case LT :
                  emitRO("SUB",ac,ac1,ac,"op<");
                  emitRM("JLT",ac,2,pc,"br if true");
                  emitRM("LDC",ac,0,ac,"false case");
                  emitRM("LDA",pc,1,pc,"unconditional jmp");
                  emitRM("LDC",ac,1,ac,"true case");
                  break;
              case EQ :
                  emitRO("SUB",ac,ac1,ac,"op==");
                  emitRM("JEQ",ac,2,pc,"br if true");
                  emitRM("LDC",ac,0,ac,"false case");
                  emitRM("LDA",pc,1,pc,"unconditional jmp");
                  emitRM("LDC",ac,1,ac,"true case");
                  break;
              default:
                  emitComment("BUG: Unknown operator");
                  break;
          } /* case op */
          if (TraceCode) emitComment("<- Op");
          break; /* OpK */

      default:
          break;
      }
} /* genExp */

/* Procedure cGen recursively generates code by
 * tree traversal
 */
static void cGen(TreeNode * tree)
{
    if (tree!=NULL)
    {
        switch (tree->nodekind) {
          case StmtK:
            genStmt(tree);
            break;
          case ExpK:
            genExp(tree);
            break;
          default:
            break;
        }
```

```c
      cGen(tree->sibling);
   }
}


/* * * * * * * * * * * * * * * * * * * * * * * * * * */
/* the primary function of the code generator * /
/* * * * * * * * * * * * * * * * * * * * * * * * * * */
/* Procedure codeGen generates code to a code
 * file by traversal of the syntax tree. The
 * second parameter (codefile) is the file name
 * of the code file, and is used to print the
 * file name as a comment in the code file
 */
void codeGen(TreeNode * syntaxTree, char * codefile)
{
    char * s=malloc(strlen(codefile)+7);
    strcpy(s,"File: ");
    strcat(s,codefile);
    emitComment("TINY Compilation to TM Code");
    emitComment(s);
    /* generate standard prelude */
    emitComment("Standard prelude:");
    emitRM("LD",mp,0,ac,"load maxaddress from location 0");
    emitRM("ST",ac,0,ac,"clear location 0");
    emitComment("End of standard prelude.");
    /* generate code for TINY program */
    cGen(syntaxTree);
    /* finish */
    emitComment("End of execution.");
    emitRO("HALT",0,0,0,"");
}
```

源代码清单 tiny-17: sample.txt

```
{  Sample program
   int TINY language-
   computes factorial
}
read x; { input an integer }
if 0<x then { don't compute if x<=0 }
  fact :=1;
```

```
    repeat
      fact :=fact * x;
      x :=x-1
    until
      x=0;
    write fact { output factorial of x }
end
```

三、TM 机

TM 机的汇编语言作为 TINY 编译器的目标语言。TM 机的指令仅够作为诸如 TINY 这样的小型语言的目标。实际上 TM 具有精简指令集计算机（RISC）的一些特性。在 RISC 中，所有的算法和测试均须在寄存器中进行，而且地址，地址模式极为有限。为了使读者了解到该机制的简便之处，我们将下面的 C 表达式的代码

```
    a[index]=6
```

翻译成 TM 汇编语言

```
    LDC 1, 0 (0)        load 0 into reg 1
```

下面的指令假设 index 在存储器地址 10 中

```
    LD 0, 10 (1)        load val at (10+R1) into R0
    LDC 1, 2, (0)       load 2 into reg 1
    MUL 0, 1, 0         put R1 * R0 into R0
    LDC 1, 0, (0)       load 0 into reg 1
```

下面的指令假设 a 在存储器地址 20 中

```
    LDA 1, 20, (1)      load 20+R1 into R0
    ADD 0, 1, 0         put R1+R0 into R0
    LDC 1, 6, (0)       load 6 into reg 1
    ST 1, 0, (0)        store R1 at 0+R0
```

装入操作中有 3 个地址模式并且是由不同的指令给出的：LDC 是"装入常量"LD 是"由存储器装入"，而 LDA 是"装入地址"。另外，该地址通常必须给成"寄存器+偏差"值。例如"10(1)"，它代表将偏差 10 加到寄存器 1 的内容中计算该地址（因为在前面的指令中，0 已经被装入到寄存器 1 中，这实际是指绝对位置 10）。我们还看到算数指令 MUL 和 ADD 可以是"三元"指令且只有寄存器操作数，其中可以单独确定结果的目标寄存器。

TM 机的模拟程序直接从一个文件中读取汇编代码并执行它，因此应避免将由汇编语言翻译为机器代码的过程复杂化。但是，这个模拟程序并非一个真正的汇编程序，它并没有符号地址或标号。因此，TINY 编译器必须仍然计算跳转的绝对地址。此外为了避

免与外部的输入/输出例程连接的复杂性，TM 机有内部整型的 I/O 设备；在模拟时，它们都对标准设备读写。

CP Lab 可以将 tm.c（参见源代码清单 tm-1）源代码编译成 TM 模拟程序，并运行 TM 指令。具体的操作可以遵循以下的步骤：

(1) 打开 CP Lab 在"文件"菜单中选择"新建"，然后单击"项目"，打开"新建项目"对话框。使用模板"TM 机"新建一个项目。

(2) 在"项目管理器"窗口中，双击 sample.tm.txt 文件（参见源代码清单 tm-2）可以看到这个文件就是 TINY 编译器由 sample.txt 源文件生成的目标代码文件。

(3) 按 F7 键生成项目。

(4) 按 F5 键启动调试，CP Lab 会自动启动 TM 机执行 sample.tm.txt 文件中的程序。按以下计算可以得到 7 的阶乘：

```
TM simulation(enter h for help)…
Enter command: go
Enter value for IN instruction: 7
OUT instruction prints: 5040
HALT: 0, 0, 0
Halted
Enter command: quit
Simulation done.
```

源代码清单 tm-1：tm.c

```
/****************************/
/* File: tm.c               */
/* The TM ("Tiny Machine") computer     */
/* Compiler Construction: Principles and Practice  */
/* Kenneth C. Louden        */
/****************************/

#include<stdio.h>
#include<stdlib.h>
#include<string.h>
#include<ctype.h>

#ifndef   TRUE
#define   TRUE 1
#endif
#ifndef   FALSE
#define   FALSE 0
#endif
```

```
/ * * * * * * * const * * * * * * * /
#define  IADDR_SIZE 1024 /* increase for large programs */
#define  DADDR_SIZE 1024 /* increase for large programs */
#define  NO_REGS 8
#define  PC_REG 7

#define  LINESIZE 121
#define  WORDSIZE 20

/ * * * * * * * type * * * * * * * /

typedef enum {
    opclRR,              /* reg operands r,s,t */
    opclRM,              /* reg r, mem d+s */
    opclRA               /* reg r, int d+s */
    } OPCLASS;

typedef enum {
    /* RR instructions */
    opHALT,              /* RR    halt, operands are ignored */
    opIN,                /* RR    read into reg(r); s and t are ignored */
    opOUT,               /* RR    write from reg(r), s and t are ignored */
    opADD,               /* RR    reg(r)=reg(s)+reg(t) */
    opSUB,               /* RR    reg(r)=reg(s)-reg(t) */
    opMUL,               /* RR    reg(r)=reg(s)*reg(t) */
    opDIV,               /* RR    reg(r)=reg(s)/reg(t) */
    opRRLim,             /* limit of RR opcodes */

    /* RM instructions */
    opLD,                /* RM    reg(r)=mem(d+reg(s)) */
    opST,                /* RM    mem(d+reg(s))=reg(r) */
    opRMLim,             /* Limit of RM opcodes */

    /* RA instructions */
    opLDA,               /* RA    reg(r)=d+reg(s) */
    opLDC,               /* RA    reg(r)=d; reg(s) is ignored */
    opJLT,               /* RA    if reg(r)<0 then reg(7)=d+reg(s) */
    opJLE,               /* RA    if reg(r)<=0 then reg(7)=d+reg(s) */
    opJGT,               /* RA    if reg(r)>0 then reg(7)=d+reg(s) */
    opJGE,               /* RA    if reg(r)>=0 then reg(7)=d+reg(s) */
    opJEQ,               /* RA    if reg(r)==0 then reg(7)=d+reg(s) */
```

```
        opJNE,       /* RA     if reg(r)!=0 then reg(7)=d+reg(s) */
        opRALim      /* Limit of RA opcodes */
     } OPCODE;

typedef enum {
    srOKAY,
    srHALT,
    srIMEM_ERR,
    srDMEM_ERR,
    srZERODIVIDE
    } STEPRESULT;

typedef struct {
     int iop ;
     int iarg1 ;
     int iarg2 ;
     int iarg3 ;
    } INSTRUCTION;

/* * * * * * * * * vars * * * * * * * * */
int iloc=0;
int dloc=0;
int traceflag=FALSE;
int icountflag=FALSE;

INSTRUCTION iMem [IADDR_SIZE];
int dMem [DADDR_SIZE];
int reg [NO_REGS];

char * opCodeTab[]
     ={"HALT","IN","OUT","ADD","SUB","MUL","DIV","????",
        /* RR opcodes */
        "LD","ST","????", /* RM opcodes */
        "LDA","LDC","JLT","JLE","JGT","JGE","JEQ","JNE","????"
        /* RA opcodes */
        };

char * stepResultTab[]
     ={"OK","Halted","Instruction Memory Fault",
        "Data Memory Fault","Division by 0"
        };

char pgmName[20];
FILE * pgm ;
```

```c
char in_Line[LINESIZE];
int lineLen;
int inCol ;
int num ;
char word[WORDSIZE];
char ch ;
int done ;

/***************************/
int opClass(int c)
{   if        (c<=opRRLim) return (opclRR);
    else if   (c<=opRMLim) return (opclRM);
    else                   return (opclRA);
} /* opClass */

/***************************/
void writeInstruction (int loc)
{   printf("%5d: ", loc);
    if ((loc>=0) && (loc<IADDR_SIZE))
    {   printf("%6s%3d,", opCodeTab[iMem[loc].iop], iMem[loc].iarg1);
        switch (opClass(iMem[loc].iop))
        {   case opclRR: printf("%1d,%1d", iMem[loc].iarg2, iMem[loc].iarg3);
                     break;
            case opclRM:
            case opclRA: printf("%3d(%1d)", iMem[loc].iarg2, iMem[loc].iarg3);
                     break;
        }
        printf ("\n");
    }
} /* writeInstruction */

/***************************/
void getCh (void)
{   if (++inCol<lineLen)
     ch=in_Line[inCol];
    else ch=' ';
} /* getCh */

/***************************/
int nonBlank (void)
{   while ((inCol<lineLen)
         && (in_Line[inCol]==' '))
      inCol++;
```

```
    if (inCol<lineLen)
    { ch=in_Line[inCol];
      return TRUE; }
    else
    { ch=' ';
      return FALSE; }
} /* nonBlank */

/* * * * * * * * * * * * * * * * * * * * * * * * * * * */
int getNum (void)
{   int sign;
    int term;
    int temp=FALSE;
    num=0;
    do
    {  sign=1;
          while (nonBlank() && ((ch=='+') || (ch=='-')))
       {  temp=FALSE;
          if (ch=='-') sign=-sign;
          getCh();
       }
       term=0;
       nonBlank();
       while (isdigit(ch))
       { temp=TRUE;
         term=term * 10 + (ch-'0');
         getCh();
       }
        num=num + (term * sign);
    } while ((nonBlank()) && ((ch=='+') || (ch=='-')));
     return temp;
   } /* getNum */

/* * * * * * * * * * * * * * * * * * * * * * * * * * */
int getWord (void)
{   int temp=FALSE;
    int length=0;
    if (nonBlank ())
    { while (isalnum(ch))
       {  if (length<WORDSIZE-1) word [length++]=ch;
          getCh();
       }
       word[length]='\0';
       temp= (length!=0);
```

```c
      }
      return temp;
} /* getWord */

/* * * * * * * * * * * * * * * * * * * * * * * * * * */
int skipCh (char c )
{   int temp=FALSE;
    if (nonBlank() && (ch==c))
    {   getCh();
        temp=TRUE;
    }
    return temp;
} /* skipCh */

/* * * * * * * * * * * * * * * * * * * * * * * * * * */
int atEOL(void)
{   return (! nonBlank ());
} /* atEOL */

/* * * * * * * * * * * * * * * * * * * * * * * * * * */
int error(char *msg, int lineNo, int instNo)
{   printf("Line %d",lineNo);
    if (instNo>=0) printf(" (Instruction %d)",instNo);
    printf("  %s\n",msg);
    return FALSE;
} /* error */

/* * * * * * * * * * * * * * * * * * * * * * * * * * */
int readInstructions (void)
{   OPCODE op;
    int arg1, arg2, arg3;
    int loc, regNo, lineNo;
    for (regNo=0; regNo<NO_REGS; regNo++)
        reg[regNo]=0;
    dMem[0]=DADDR_SIZE-1;
    for (loc=1; loc<DADDR_SIZE; loc++)
        dMem[loc]=0;
    for (loc=0; loc<IADDR_SIZE; loc++)
    {   iMem[loc].iop=opHALT;
        iMem[loc].iarg1=0;
        iMem[loc].iarg2=0;
        iMem[loc].iarg3=0;
    }
```

```
   lineNo=0;
while (! feof(pgm))
{  fgets(in_Line, LINESIZE-2, pgm );
   inCol=0;
   lineNo++;
   lineLen=strlen(in_Line)-1;
   if (in_Line[lineLen]=='\n') in_Line[lineLen]='\0';
   else in_Line[++lineLen]='\0';
   if ((nonBlank()) && (in_Line[inCol]!='*'))
   {  if (! getNum())
         return error("Bad location", lineNo,-1);
      loc=num;
      if (loc>IADDR_SIZE)
         return error("Location too large",lineNo,loc);
      if (! skipCh(':'))
         return error("Missing colon", lineNo,loc);
      if (! getWord ())
         return error("Missing opcode", lineNo,loc);
       op=opHALT;
      while ((op<opRALim)
            && (strncmp(opCodeTab[op], word, 4)!=0))
         op++;
      if (strncmp(opCodeTab[op], word, 4)!=0)
         return error("Illegal opcode", lineNo,loc);
      switch (opClass(op))
        { case opclRR :
  /* * * * * * * * * * * * * * * * * * * * * * * * * */
      if ((! getNum ()) || (num<0) || (num>=NO_REGS))
          return error("Bad first register", lineNo,loc);
      arg1=num;
      if (! skipCh(','))
          return error("Missing comma", lineNo, loc);
      if ((! getNum ()) || (num<0) || (num>=NO_REGS))
          return error("Bad second register", lineNo, loc);
      arg2=num;
      if (! skipCh(','))
          return error("Missing comma", lineNo,loc);
      if ((! getNum ()) || (num<0) || (num>=NO_REGS))
          return error("Bad third register", lineNo,loc);
      arg3=num;
      break;

      case opclRM :
      case opclRA :
```

```c
            /* * * * * * * * * * * * * * * * * * * * * * * * * */
        if ((! getNum ()) || (num< 0) || (num>=NO_REGS))
            return error("Bad first register", lineNo,loc);
        arg1=num;
        if (! skipCh(','))
            return error("Missing comma", lineNo,loc);
        if (! getNum ())
            return error("Bad displacement", lineNo,loc);
        arg2=num;
        if (! skipCh('(') &&! skipCh(','))
            return error("Missing LParen", lineNo,loc);
        if ((! getNum ()) || (num< 0) || (num>=NO_REGS))
            return error("Bad second register", lineNo,loc);
        arg3=num;
        break;
        }
        iMem[loc].iop=op;
        iMem[loc].iarg1=arg1;
        iMem[loc].iarg2=arg2;
        iMem[loc].iarg3=arg3;
            }
        }
        return TRUE;
} /* readInstructions */

/* * * * * * * * * * * * * * * * * * * * * * * * * */
STEPRESULT stepTM (void)
{   INSTRUCTION currentinstruction ;
    int pc ;
    int r,s,t,m ;
    int ok;

    pc=reg[PC_REG];
    if ((pc< 0) || (pc>IADDR_SIZE) )
        return srIMEM_ERR;
    reg[PC_REG]=pc +1;
    currentinstruction=iMem[ pc ];
    switch (opClass(currentinstruction.iop))
    { case opclRR :
/* * * * * * * * * * * * * * * * * * * * * * * * * */
      r=currentinstruction.iarg1;
      s=currentinstruction.iarg2;
      t=currentinstruction.iarg3;
      break;
```

```c
      case opclRM :
/* * * * * * * * * * * * * * * * * * * * * * * * * * * */
    r=currentinstruction.iarg1;
    s=currentinstruction.iarg3;
    m=currentinstruction.iarg2+reg[s];
    if((m<0) || (m>DADDR_SIZE))
    return srDMEM_ERR;
    break;

    case opclRA :
/* * * * * * * * * * * * * * * * * * * * * * * * * * * * * * * * */
    r=currentinstruction.iarg1;
    s=currentinstruction.iarg3;
    m=currentinstruction.iarg2+reg[s];
    break;
} /* case */

switch (currentinstruction.iop)
{  /* RR instructions */
    case opHALT :
/* * * * * * * * * * * * * * * * * * * * * * * * * * * * * * * */
    printf("HALT: %1d,%1d,%1d\n",r,s,t);   return srHALT;
    /* break; */

case opIN :
/* * * * * * * * * * * * * * * * * * * * * * * * * * * * */
    do
     { printf("Enter value for IN instruction: ");
        fflush (stdin);
        fflush (stdout);
        gets(in_Line);
        lineLen=strlen(in_Line);
        inCol=0;
        ok=getNum();
        if (! ok) printf ("Illegal value\n");
        else reg[r]=num;
     }
    while (! ok);
    break;

    case opOUT :
      printf ("OUT instruction prints: %d\n", reg[r]);
      break;
```

```c
        case opADD :    reg[r]=reg[s]+reg[t]; break;
        case opSUB :    reg[r]=reg[s]-reg[t]; break;
        case opMUL :    reg[r]=reg[s]*reg[t]; break;

        case opDIV :
        /********************************************/
          if (reg[t]!=0) reg[r]=reg[s]/ reg[t];
          else return srZERODIVIDE;
          break;

        /************RM instructions *******************/
        case opLD :     reg[r]=dMem[m]; break;
        case opST :     dMem[m]=reg[r]; break;

        /************RA instructions *******************/
        case opLDA :    reg[r]=m; break;
        case opLDC :    reg[r]=currentinstruction.iarg2;    break;
        case opJLT :    if (reg[r]<0) reg[PC_REG]=m;        break;
        case opJLE :    if (reg[r]<=0) reg[PC_REG]=m;       break;
        case opJGT :    if (reg[r]>0) reg[PC_REG]=m;        break;
        case opJGE :    if (reg[r]>=0) reg[PC_REG]=m;       break;
        case opJEQ :    if (reg[r]==0) reg[PC_REG]=m;       break;
        case opJNE :    if (reg[r]!=0) reg[PC_REG]=m;       break;

        /* end of legal instructions */
} /* case */
return srOKAY;
} /* stepTM */

/*********************************************/
int doCommand (void)
{   char cmd;
    int stepcnt=0, i;
    int printcnt;
    int stepResult;
    int regNo, loc;
    do
    {   printf ("Enter command: ");
        fflush (stdin);
        fflush (stdout);
```

```
      gets(in_Line);
      lineLen=strlen(in_Line);
      inCol=0;
   }
   while (! getWord ());

   cmd=word[0];
   switch (cmd)
   {  case 't' :
   /* * * * * * * * * * * * * * * * * * * * * * * * * * * * * * * * * */
         traceflag=! traceflag;
         printf("Tracing now ");
         if (traceflag) printf("on.\n"); else printf("off.\n");
         break;

      case 'h' :
   /* x x x x x x x x x x x x x x x x x x x x x x x x x x x x x x x x */
         printf("Commands are:\n");
         printf("   s(tep<n>      "\
                "Execute n (default 1) TM instructions\n");
         printf("   g(o           "\
                "Execute TM instructions until HALT\n");
         printf("   r(egs         "\
                "Print the contents of the registers\n");
         printf("   i(Mem<b<n>>"\
                "Print n iMem locations starting at b\n");
         printf("   d(Mem<b<n>>"\
                "Print n dMem locations starting at b\n");
         printf("   t(race        "\
                "Toggle instruction trace\n");
         printf("   p(rint        "\
                "Toggle print of total instructions executed"\
                " ('go' only)\n");
         printf("   c(lear        "\
                "Reset simulator for new execution of program\n");
         printf("   h(elp         "\
                "Cause this list of commands to be printed\n");
         printf("   q(uit         "\
                "Terminate the simulation\n");
         break;
```

```c
        case 'p' :
/* * * * * * * * * * * * * * * * * * * * * * * * * * * * * * * * */
          icountflag=!icountflag;
          printf("Printing instruction count now ");
          if (icountflag) printf("on.\n"); else printf("off.\n");
          break;

        case 's' :
/* * * * * * * * * * * * * * * * * * * * * * * * * * * * * * * * */
          if (atEOL ()) stepcnt=1;
          else if (getNum ()) stepcnt=abs(num);
          else    printf("Step count?\n");
          break;

        case 'g' :   stepcnt=1;    break;

        case 'r' :
/* * * * * * * * * * * * * * * * * * * * * * * * * * * * * * * * */
          for (i=0; i<NO_REGS; i++)
          { printf("%1d: %4d    ", i,reg[i]);
            if ((i%4)==3) printf ("\n");
          }
          break;

        case 'i' :
/* * * * * * * * * * * * * * * * * * * * * * * * * * * * * * * * */
          printcnt=1;
          if (getNum ())
          { iloc=num;
            if (getNum ()) printcnt=num;
          }
          if (! atEOL ())
            printf ("Instruction locations?\n");
          else
          { while ((iloc>=0) && (iloc<IADDR_SIZE)
                  && (printcnt>0))
              { writeInstruction(iloc);
                iloc++;
                printcnt--;
              }
          }
          break;
```

```
    case 'd' :
/ * * * * * * * * * * * * * * * * * * * * * * * * * * * * * * * * * /
  printcnt=1;
  if (getNum ())
  { dloc=num;
    if (getNum ()) printcnt=num;
  }
  if (! atEOL ())
    printf("Data locations?\n");
  else
  { while ((dloc>=0) && (dloc<DADDR_SIZE)
          && (printcnt>0))
    { printf("%5d: %5d\n",dloc,dMem[dloc]);
      dloc++;
      printcnt--;
    }
  }
  break;

case 'c' :
/ * * * * * * * * * * * * * * * * * * * * * * * * * * * * * * * * /
  iloc=0;
  dloc=0;
  stepcnt=0;
  for (regNo=0; regNo<NO_REGS; regNo++)
      reg[regNo]=0;
  dMem[0]=DADDR_SIZE-1;
  for (loc=1; loc<DADDR_SIZE; loc++)
      dMem[loc]=0;
  break;

case 'q' : return FALSE; /* break; */

default : printf("Command %c unknown.\n", cmd); break;
} /* case */
stepResult=srOKAY;
if (stepcnt>0)
{  if (cmd=='g')
      { stepcnt=0;
        while (stepResult==srOKAY)
        { iloc=reg[PC_REG];
          if (traceflag) writeInstruction(iloc);
          stepResult=stepTM ();
```

```
                    stepcnt++;}
              if (icountflag)
         printf("Number of instructions executed=%d\n",stepcnt);
       }
     else
     { while ((stepcnt>0) && (stepResult==srOKAY))
         { iloc=reg[PC_REG];
           if (traceflag) writeInstruction(iloc);
           stepResult=stepTM ();
           stepcnt--;
         }
      }
      printf("%s\n",stepResultTab[stepResult]);
     }
     return TRUE;
   } /* doCommand */

/* * * * * * * * * * * * * * * * * * * * * * * * * * * * * * * * * * /
/* E X E C U T I O N    B E G I N S    H E R E */
/* * * * * * * * * * * * * * * * * * * * * * * * * * * * * * * * * * /

main(int argc, char * argv[])
{  if (argc!=2)
   { printf("usage: %s<filename>\n",argv[0]);
     exit(1);
   }
   strcpy(pgmName,argv[1]);
   if (strchr (pgmName, '.')==NULL)
      strcat (pgmName,".tm");
   pgm=fopen(pgmName,"r");
   if (pgm==NULL)
   { printf("file '%s' not found\n",pgmName);
     exit(1);
   }

   /* read the program */
   if (! readInstructions ())
       exit(1);
   /* switch input file to terminal */
   /* reset(input); */
   /* read-eval-print */
   printf("TM simulation (enter h for help)...\n");
   do
```

```
        done=! doCommand ();
while (! done);
printf("Simulation done.\n");

return 0;
}
```

源代码清单 tm-2: sample.tm.txt

```
0:     LD    6,0(0)
1:     ST    0,0(0)
2:     IN    0,0,0
3:     ST    0,0(5)
4:     LDC   0,0(0)
5:     ST    0,0(6)
6:     LD    0,0(5)
7:     LD    1,0(6)
8:     SUB   0,1,0
9:     JLT   0,2(7)
10:    LDC   0,0(0)
11:    LDA   7,1(7)
12:    LDC   0,1(0)
14:    LDC   0,1(0)
15:    ST    0,1(5)
16:    LD    0,1(5)
17:    ST    0,0(6)
18:    LD    0,0(5)
19:    LD    1,0(6)
20:    MUL   0,1,0
21:    ST    0,1(5)
22:    LD    0,0(5)
23:    ST    0,0(6)
24:    LDC   0,1(0)
25:    LD    1,0(6)
26:    SUB   0,1,0
27:    ST    0,0(5)
28:    LD    0,0(5)
29:    ST    0,0(6)
30:    LDC   0,0(0)
31:    LD    1,0(6)
32:    SUB   0,1,0
33:    JEQ   0,2(7)
```

```
34:    LDC    0,0(0)
35:    LDA    7,1(7)
36:    LDC    0,1(0)
37:    JEQ    0,-22(7)
38:    LD     0,1(5)
39:    OUT    0,0,0
13:    JEQ    0,27(7)
40:    LDA    7,0(7)
41:    HALT   0,0,0
```

参 考 文 献

[1] Kenneth C Louden. Compiler Construction: Principles and Practice. 冯博琴,冯岚,等译. 编译原理及实践. 北京: 机械工业出版社,2009.
[2] 陈火旺,刘春林,等. 程序设计语言编译原理. 3版. 北京: 国防工业出版社,2011.
[3] 张菁. 编译原理及实践(中英双语版). 北京: 清华大学出版社,2007.
[4] Andrew W Appel. 现代编译原理C语言描述. 北京: 人民邮电出版社,2006.
[5] 王雷,刘志成,周晶. 编译原理课程设计. 北京: 机械工业出版社,2005.
[6] 王晓斌. 程序设计语言与编译. 北京: 电子工业出版社,2015.
[7] 伍春香. 编译原理——习题与解析. 北京: 清华大学出版社,2006.
[8] 赵炯. Linux内核完全剖析. 北京: 机械工业出版社,2006.
[9] [英] Peter Abel 著. IBM PC 汇编语言程序设计. 5版. 沈美明,温冬婵译. 北京: 人民邮电出版社,2002.
[10] 刘星,等. 计算机接口技术. 北京: 机械工业出版社,2003.
[11] 唐朔飞. 计算机组成原理. 北京: 高等教育出版社,2000.
[12] [希腊] Diomidis Spinellis 著. 代码阅读方法与实践. 赵学良译. 北京: 清华大学出版社,2004.
[13] [美] 科学、工程和公共政策委员会著. 怎样当一名科学家. 何传启译. 北京: 科学出版社,1996.
[14] CP Lab 帮助文档. 北京英真时代科技有限公司. http://www.engintime.com/node/27.
[15] Intel Co. INTEL 80386 Programmes's Refercence Manual 1986, INTEL CORPORATION,1987.
[16] Intel Co. IA-32 Intel Architecture Software Developer's Manual Volume. 3: System Programming Guide. http://www.intel.com/,2005.
[17] Microsoft Co. FAT: General Overview of On-Disk Format. MICROSOFT CORPORATION,1999.
[18] IEEE-CS,ACM. Computing Curricula 2001 Computer Science,2001.
[19] The NASM Development Team. NASM-The Netwide Assembler. Version 2.04,2008.
[20] Bochs simulation system. http://bochs.sourceforge.net/.